Applied Mathematical Sciences | Volume 36

To Grace Wahba,
 with very best wishes,
  Michael Ghil

Dynamic Meteorology: Data Assimilation Methods

Lennart Bengtsson
Michael Ghil
Erland Källén, editors

Springer-Verlag
New York   Heidelberg   Berlin

Lennart Bengtsson
European Centre for
Medium Range Weather
Forecasts
Shinfield Park
Reading, Berkshire
England RG2 9AX

Michael Ghil
Courant Institute
New York University
251 Mercer Street
New York, New York
10012

Erland Källén
European Centre for
Medium Range Weather
Forecasts
Shinfield Park
Reading, Berkshire
England RG2 9AX

Library of Congress Cataloging in Publication Data

Dynamic Methodology: Data Assimilation Methods.

(Applied mathematical sciences; v. 36)
Contents: An overview of meteorological data assimilation / P. Morel—A review of methods for objective analysis / Nils Gustavsson—Normal mode initialization / R. Daley—[etc.]
1. Meteorology—Data processing—Congresses.
2. Dynamic meteorology—Congresses. I. Ghil, Michael.
II. Källén, E. III. Bengtsson, L. IV. Title.
V. European Centre for Medium Range Weather Forecasts. VI. Series: Applied mathematical sciences (Springer-Verlag New York Inc.); v. 36.
QA1.A647 vol. 36 [QC874.3] 510s [551.5'028'5]81-14410
AACR2

All rights reserved. No part of this book may be translated or reproduced in any form without written permission from Springer-Verlag, 175 Fifth Avenue New York, New York 10010, U.S.A.

The use of general descriptive names, trade names, trademarks, etc. in this publication, even if the former are not especially identified, is not to be taken as a sign that such names, as understood by the Trade Marks and Merchandise Marks Act, may accordingly be used freely by anyone.

© 1981 by Springer-Verlag New York, Inc.

Printed in the United States of America.

9 8 7 6 5 4 3 2 1

ISBN 0-387-90632-0 Springer-Verlag   New York   Heidelberg   Berlin
ISBN 3-540-90632-0 Springer-Verlag   Berlin   Heidelberg   New York

PREFACE

One of the main reasons we cannot tell what the weather will be tomorrow is that we do not know accurately enough what the weather is today. Mathematically speaking, numerical weather prediction (NWP) is an initial-value problem for a system of nonlinear partial differential equations in which the necessary initial values are known only incompletely and inaccurately. Data at the initial time of a numerical forecast can be supplemented, however, by observations of the atmosphere over a time interval preceding it.

New observing systems, in particular polar-orbiting and geostationary satellites, which are providing observations continuously in time, make is absolutely necessary to find new and more satisfactory methods of assimilating meteorological observations - for the dual purpose of defining atmospheric states and of issuing forecasts from the states thus defined. Fundamental progress in this area has been made in recent years and this book attempts to give a review and some suggestions for further improvements in the field of meteorological data assimilation methods.

The European Centre for Medium Range Weather Forecasts (ECMWF) every year organises seminars for the benefit of meteorologists and geophysicists of the ECMWF Member States. The 1980 Seminar was devoted to data assimilation methods, and this book contains selected lectures from that seminar. The purpose of the seminar was twofold: it was intended to give a basic introduction to the subject, as well as an overview of the latest developments in the field.

The specific reason for devoting the 1980 Seminar to data assimilation is ECMWF's current involvement in analysing data from the Global Weather Experiment. The Global Atmospheric Research Program (GARP) was organized in 1967 by the World Meteorological Organization and the International Council of Scientific Unions to increase our scientific understanding of the Earth's atmosphere and our ability to predict its behaviour. GARP's Global Weather Experiment is the most extensive and most costly joint international research programme ever undertaken. It involved an intense one year period (1 December 1978 - 30 November 1979) of measuring various properties of the Earth's atmosphere, with the aid of observations taken from ships, aircraft, satellites, balloons and radiosondes, in addition to conventional observations from the permanent meteorological network (see Fig. 1). All these data were collected and processed by a number of meteorological Centres all over the world. The data will be used for scientific investigations in dynamic meteorology and related geophysical disciplines over the next decade.

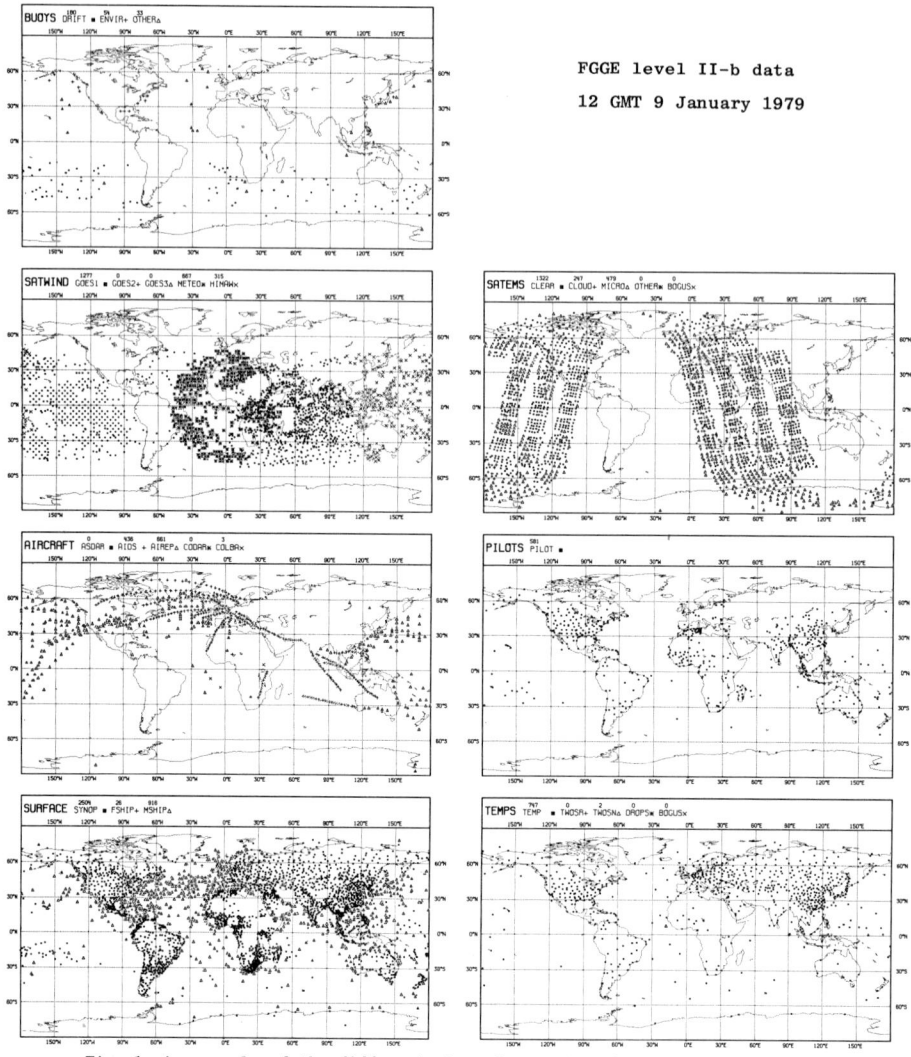

FGGE level II-b data
12 GMT 9 January 1979

Fig. 1   An example of the different observing systems in use during the Global Weather Experiment.

Abbreviations used:

Aireps	Standard wind observations from aircraft
Asdars, Aids	High quality wind observations from aircraft
Buoys	Surface pressure observations from drifting buoys
Colba	Constant level balloons
Drops	Radiosondes dropped from aircraft
Pilots	Wind measurements from ascending balloons
Satems	Temperature measurements from polar orbiting satellites
Satwind	Cloud drift wind measurements from geostationary satellites
Ships	Surface observations from ships
Synops	Surface observations from land
Temps	Temperature, humidity and wind measurements from radiosondes

The volume opens with a general review by P. Morel. This is followed by a careful summary of methods presently used by numerical prediction centers in the objective analysis of time-synchronous observations, written by N. Gustafsson.

The equations of motion used in NWP admit two types of waves: slow Rossby waves and fast inertia-gravity waves. An initial distribution of energy between these waves similar to that in the atmosphere is achieved in numerical models by initialization procedures. The recently developed method for nonlinear normal mode initialization is covered in detail by R. Daley. Mathematical aspects of the four-dimensional assimilation of data in space and time are presented by O. Talagrand.

An alternative approach to the initialization problem and a demonstration of how methods taken from estimation theory may be applied to data assimilation problems are given by M. Ghil. Some examples from operational data assimilation carried out at ECMWF in relation to the Global Weather Experiment are given by M. Kanamitsu. A full summary of all papers given during the seminar has been published by ECMWF. In this volume, only papers of a general theoretical interest have been included. It is hoped that the volume can serve as a self-contained introduction to a rapidly expanding area of research, which raises many interesting geophysical and mathematical problems.

As an Appendix we have included A. Eliassen's classical 1954 paper which presented one of the first attempts to design an objective, automated method to analyse meteorological data and was a precursor of current 'optimum interpolation' methods. We gratefully acknowledge the permission from the author to reprint it here.

It is a pleasure to acknowledge all those who have been involved in typing the manuscript and drafting the figures which are included in this book. Most of all, we thank all of the lecturers and participants for contributing to the scientific success of the 1980 ECMWF Seminar.

The Editors, Reading and New York, 1981.

C O N T E N T S   Page

List of authors   1

List of participants   2

AN OVERVIEW OF METEOROLOGICAL DATA ASSIMILATION   5
P. Morel

A REVIEW OF METHODS FOR OBJECTIVE ANALYSIS   17
Nils Gustavsson

NORMAL MODE INITIALIZATION   77
R. Daley

ASSIMILATION OF ASYNOPTIC DATA AND THE INITIALIZATION PROBLEM   111
K.P. Bube and M. Ghil

APPLICATIONS OF ESTIMATION THEORY TO NUMERICAL WEATHER PREDICTION   139
M. Ghil, S. Cohn, J. Tavantzis, K. Bube, E. Isaacson

CONVERGENCE OF ASSIMILATION PROCEDURES   225
O. Talagrand

SOME CLIMATOLOGICAL AND ENERGY BUDGET CALCULATIONS USING   263
THE FGGE III-b ANALYSES DURING JANUARY 1979
M. Kanamitsu

APPENDIX   319
PROVISIONAL REPORT ON CALCULATION OF SPATIAL COVARIANCE AND
AUTOCORRELATION OF THE PRESSURE FIELD
A. Eliassen

LIST OF AUTHORS (Present at the Seminar)
with present affiliations

R. DALEY
National Center for Atmospheric Research
P.O. Box 3000
Boulder, Colorado 80307
USA

M. GHIL
Courant Institute of Mathematical Sciences
251 Mercer Street
New York, N.Y. 10012
USA

N. GUSTAVSSON
Swedish Meteorological and Hydrological Institute
Box 923
S-60119, Norrköping
Sweden

M. KANAMITSU
Electronic Computation Centre
Japan Meteorological Agency
2-3-4, Otemachi-Chiyoda-ku
Tokyo 100, Japan

P. MOREL
Centre National d'Études Spatiales
129 Rue de l'Université
F-750 07 Paris
France

O. TALAGRAND
Laboratoire de Meteorologie Dynamique
C.N.R.S.
24 rue Lhomond
F-75231 Paris
France

LIST OF PARTICIPANTS

M. Piringer
Zentralanstalt für Meteorologie u. Geodynamik,
Wien, Austria

J. Vanderborght
Institut Royal Météorologique de Belgique
Bruxelles, Belgium

A. Walløe Hansen
Danish Meteorological Institute
København, Denmark

A.M. Jørgensen
Danish Meteorological Institute
København, Denmark

K. Eerola
Finnish Meteorological Institute
Helsinki, Finland

S. Brière
Direction de la Météorologie
Boulogne, France

Y. Durand
Météorologie Nationale
Paris, France

J.P. Pailleux
Météorologie Nationale
Paris, France

J.P. Volmer
Direction de la Météorologie
Boulogne, France

D. Dedenbach
Amt für Wehrgeophysik
Traben-Trarbach
FR Germany

St. Emeis
Met. Inst. Universität
Bonn, FR Germany

M. Hantel
Met. Inst. Universität
Bonn, FR Germany

A. Kaestner
Deutscher Wetterdienst
Offenbach, FR Germany

J.-U. Schwirner
Deutscher Wetterdienst
Offenbach, FR Germany

P. Speth
Inst. f. Geophysik u. Meteorologie
Universität zu Köln, FR Germany

G. Barbounakis
National Meteorological Service
Athens, Greece

G.A. Dalu
IFA
Rome, Italy

C. Finizio
Servizio Meteorologico
Rome, Italy

R. Purini
IFA
Rome, Italy

C. Valenti
IFA
Rome, Italy

J.R. Bijlsma
KNMI
De Bilt, Netherlands

B.J. de Haan
KNMI
De Bilt, Netherlands

H. Timmerman
KNMI
De Bilt, Netherlands

K. Bjørheim
Det Norske Meteorologiske Institutt
Oslo, Norway

I. Carlsson
Swedish Royal Air Force
Stockholm, Sweden

S. Gollvik
University of Stockholm
Sweden

H. Lejenäs
University of Stockholm
Sweden

R. Lindqvist
Swedish Royal Air Force
Stockholm, Sweden

N.D. Que
SMHI
Norrköping, Sweden

Å. Johansson
University of Stockholm
Sweden

LIST OF PARTICIPANTS (Continued)

L. Thaning
University of Stockholm
Sweden

A. Ullerstig
SMHI
Norrköping, Sweden

E. Özsoy
Middle East Tech. University
Marine Science Dept.,
Icel, Turkey

M.J. Atkins
Meteorological Office
Bracknell, UK

G. Barnes
Meteorological Office
Bracknell, UK

S. Bell
Meteorological Office
Bracknell, UK

P.J. Everson
University of Exeter
Exter, UK

M.A. Pedder
University of Reading
Department of Meteorology
Reading, UK

R.J. Purser
Meteorological Office
Bracknell, UK

P.R. Julian
National Center for Atmospheric Research
Boulder, Colorado
USA

AN OVERVIEW OF METEOROLOGICAL DATA ASSIMILATION

Pierre Morel
University of Paris
Laboratoire de Meteorologie Dynamique
24 rue Lhomond, 75005 Paris
France

1. INTRODUCTION

The process of numerical weather prediction is classically viewed as an initial value problem whereby the governing equations of geophysical fluid dynamics are integrated forward from fully determined initial values of the meteorological fields at some initial time. Given the mathematical properties of the equations of motion applied to geophysical fluids and the complexity of energetic processes in the atmosphere, solving this "initial value problem" is by itself a tremendous task: many a numerical and physical "modeler" is currently engrossed in it. Still, providing the ultimate scheme for integrating the discretized equations of atmospheric dynamics would only be half the answer. For one must also attend to the problem of determining, from observations of the real atmosphere, the initial values of the many time-dependent quantities which define the state of a discrete, numerical, atmospheric model.

In the ideal physical situation, one would expect that observations would specify unambiguously one value for each and every state parameter at the selected initial time $t = 0$. Needless to say, this ideal case never materializes in actual meteorological forecasting for a variety of reasons:

(i) conventional pressure, temperature and wind observations which are taken at the so-called synoptic times, 0000 GMT and 1200 GMT, are inadequately distributed around the planet and leave severe geographical gaps where no data are available.

(ii) conventional observations are point measurements which do not provide a correct sampling of the highly variable meteorological fields: such measurements are not representative of true volume averages as required by numerical models; conventional observations are also subject to significant random instrumental errors.

(iii) remote observations from space, leading to an indirect determination of vertical temperature profiles, go a long way toward providing a homogeneous global coverage leaving no significant data gaps over a 6-to-12 hour period of time. Such observations are not synoptic, however, but distributed in space and time following the trajectory of sunsynchronous orbiting satellites. The input from such observing systems is therefore very incomplete at any one time.

(iv) remote observations of cloud motions from geostationary satellites are available at synoptic hours and provide essentially a proper horizontal sampling of the wind field in tropical regions. But the vertical resolution, limited to one or two levels, is grossly inadequate.

(v) finally, one will note that indirect observations from space always involve physical assumptions and fairly sophisticated data processing for reconstructing the meteorological state parameters from remotely measured physical quantities. These procedures have their own deficiencies, causing significant errors, both random and systematic.

One must regretfully conclude from this brief discussion that <u>single-time data</u> sets available to numerical forecasters for updating their computations are likely to remain incomplete and inaccurate, i.e., by no means sufficient to provide themselves an adequate description of the global atmosphere. Any forecasting scheme must therefore be initiated or "initialized" at time t = 0 (say) by merging the new observations with the currently estimated meteorological fields, computed on the basis of earlier observations collected at times t < 0. Formally, the problem consists in <u>optimizing</u> the generalized N - dimensional "trajectory" of the model, considered as a mechanical system with N degrees of freedom, while taking into account all available information at times t ⩽ 0, as well as the dynamical constraints between successive (model) states, specified by the governing dynamical equations. This process of merging new observational data with the ongoing integration of a numerical forecasting model is known as "data assimilation" or equivalently "4-dimensional data assimilation" in consideration of the time-space distribution of the data base.

Whether they recognize it or not, all numerical forecasters resort to some form of 4-dimensional data assimilation; not all forecasters, however, can claim having achieved an optimal 4-dimensional data assimilation scheme, or one close to optimality. On the contrary, most forecasting teams will readily admit that a large part of forecasting deficiencies are rooted in the imperfect assimilation of available data in their numerical prediction process. During recent years, the development of the Global Atmospheric Research Programme (GARP) and the increasing availability of non-synoptic remote observations from earth-orbiting satellites have fostered a growing interest in meteorological data analysis and triggered very significant advances towards developing effective "optimized" data assimilation methods. The purpose of this paper is to review briefly the underlying physical principles which are operative in this problem; it also provides an introduction to the other papers in this volume, which deal in further detail with various aspects of the problem.

2. STATISTICAL GRID-POINT ANALYSIS

Since observations seldom occur at the precise locations represented by the grid points of one particular model, some 3-dimensional interpolation scheme is a necessary step whenever new data must be merged with predicted fields, whether they are represented by grid-point values or spectral expansion coefficients.

The simplest approach consists in substituting the observed value for the corresponding meteorological parameter at the nearest grid point. This gives very poor results. The next simplest and most common approach consists in deriving the correction of some computed parameter at one particular location as a linear combination of the departures of actual observations obtained in the vicinity from computed field values, as interpolated to observation locations. This linear combination is usually made on the basis of appropriate coefficients or "weights" attributed to individual observations according to the radial distance of their location and their assumed "quality". Eliassen (1954) and Gandin (1963) have originally described how these weights may be optimally determined in order to minimize, in a statistical sense, the resulting interpolation error at each grid point, by taking into account the known spatial structure of the meteorological field. This so-called "optimal analysis scheme" was originally applicable to scalar fields only, i.e., to one variable at a time. A fairly recent generalisation has been proposed by Schlatter (1975) whereby different meteorological parameters may be simultaneously analysed in terms of grid-point values in a single multi-variate statistical interpolation process. Interpolation coefficients in such multivariate analysis schemes are computed on the basis of cross-correlation coefficients between different meteorological variables such as geopotential height and one component of the wind velocity, for example. These second-order correlations could in principle be derived directly from observations of the real meteorological fields. This is not usually done, however, because the autocorrelation statistics of the geopotential field are much better established than any other second-order moment of the multivariate field. It is current practice therefore, to compute the various statistical moments from the geopotential autocorrelation function alone, using the geostrophic approximation to link the wind variables with the geopotential field. Through this device, the multivariate analysis of wind and geopotential data forces a degree of geostrophic balance which turns out to be beneficial generally, although certainly not justified by any careful consideration of the dynamical effects. The student of 4-dimensional data assimilation should, however, be alert to the fact that excessive filtering of the ageostrophic component of the real atmosphere flow could follow when multivariate interpolation is applied in data-sparse regions and is too efficient at "reconstructing" the missing field values.

A further remark could be made regarding the spatial resolution of the grid, or equivalent spectral expansion, used for this analysis. The principle of multivariate interpolation is to induce a degree of geostrophic balance in the vicinity of data points. This objective, however, is attainable only if the model provides a large enough number of degrees of freedom within the influence area around the particular data point. Now, observations of meteorological fields indicate an effective correlation range of the order of 1000 to 1500 km. This means that a multivariate analysis scheme could be effective only when applied to a model with significantly better spatial resolution, i.e., no less than 300-500 km.

This condition is normally fulfilled by current general circulation and global forecasting models. But multivariate interpolation schemes may turn out to be ineffective when applied to the analysis of mesoscale features, for which a much smaller correlation range is indicated, while spatial resolutions are only moderately smaller. The statistical interpolation, or analysis aspect of the data assimilation problem is discussed in this volume by N. Gustafsson.

3. THE DYNAMICS OF ADJUSTMENT

The very first attempt at numerical weather prediction by Richardson (1922) called attention to the need to eliminate the spurious high frequency oscillations which result from inserting data taken from an external reference (in this instance, the real atmosphere), in the course of the computation. These oscillations, also referred to as "meteorological noise", have the character of gravity waves superimposed on the quasi-geostrophic flow which constitutes the "meteorological signal". The existence of these waves would not be too serious if they were essentially independent modes, uncoupled to the main flow. But this is not so in the fully non-linear interactive case of a primitive-equation model: gravity modes can draw energy from the main flow and grow rapidly in the course of time integration. The problem of dealing with high-frequency meteorological noise has thus become a central issue in data assimilation.

It is permissible, as a convenient educational device, to rely on the simple model of the linearized shallow-water equations describing an incompressible fluid in a rotating frame for the purpose of discussing the basic concepts of adjustment dynamics. We shall follow this traditional approach and write down the familiar set of governing equations for one component of the FOURIER expansion corresponding to wavevector $\vec{k}$:

$$\frac{\partial \psi}{\partial t} = -f\chi \quad , \quad (1)$$

$$\frac{\partial \chi}{\partial t} = f\psi - gh \quad , \quad (2)$$

$$\frac{\partial h}{\partial t} = Hk^2 \chi \quad , \quad (3)$$

Here the two scalar fields, stream function ψ and velocity potential χ, fully determine the 2-dimensional velocity field, while h is the departure of the free surface altitude from the mean level H. All three variables ψ, χ and h are first order quantities for a small perturbation of the state of rest of the fluid. One recognizes above the vorticity equation (1), the divergence equation (2) and the continuity equation (3). This set of equations admits three linearly independent solutions or <u>normal modes</u> for each wavevector $\vec{k}$, $k = |\vec{k}|$, namely:

- one geostrophically balanced mode which turns out to be stationary in this simple model.

- two high frequency modes corresponding to gravity waves propagating in the forward and backward directions, respectively.

One could also view each normal mode as an eigenvector X_i corresponding to each of the three eigenvalues $\lambda_1 = 0$, $\lambda_2 = \sqrt{gHk^2 + f^2}$ and $\lambda_3 = -\lambda_2$. If one now projects the equations (1) through (3) onto these normal modes, one obtains a set of ordinary differential equations which may be written symbolically as:

$$\dot{X} = -i \Lambda X \qquad (4)$$

where X is a column vector of normal mode expansion coefficients and Λ is a diagonal matrix, the elements of which are the three eigenvalues λ_i.

Initializing this simple dynamical model with an arbitrary state vector X at time t = 0 or equivalently, an arbitrary selection of the fields ψ, χ and h, will normally result in exciting all three normal modes about equally. In the simple linear model, the initial distribution of perturbation energy will simply be invariant at all times t > 0 without amplification nor damping.

In the more realistic non-linear dynamical models, however, the non-linear advection terms and the forcing terms have a non-zero projection on the set of normal modes X, so that equation (4) would now read:

$$X = - i \Lambda X + N (X, X) \qquad (5)$$

and allow model coupling, which in turn induces a tendency towards equipartition of energy among the three competing modes. It is a fact that, barring ad hoc precautions, numerical models develop a much larger amount of gravity-wave energy than actually found in the real atmosphere. Conversely, one could well take the view that the absence of significant gravity oscillations in free atmospheric flows is a remarkable and yet not fully explained property of the Earth's atmosphere. One possible reason for it might be found in the major energy conversion process associated with baroclinic instability, which favours the generation of eddy mechanical energy in the form of quasi-geostrophic perturbations of the mean flow. Alternately, another cause for selective dissipation of gravity oscillations might be looked for in the basic non-linear dynamics of spectral energy transfer which favours the concentration of quasi two-dimensional motions in the larger scales, while the energy of 3-dimensional gravity waves is allowed to cascade towards smaller scales and dissipate by friction. Yet another possible explanation could be seen in the dynamics of small-scale energetic processes (like moist convection) which could be selectively triggered by gravity oscillations.

Whatever, the main cause for the damping of gravity modes in the real atmosphere is, it is imperative to develop <u>filtering techniques</u> for removing the spuriously large gravity oscillations induced in numerical models by non-linear interactions, as well as by the insertion of heterogeneous data.

The simplest and historically the first approach was simply to exclude the possibility of such high-frequency oscillations by reducing the model atmosphere dynamics to the quasi-geostrophic response allowed exclusively by a "filtered system" as the balance equations. This approach so severely restricts the model dynamics, however, that corresponding forecasts are not very useful beyond 24 hours. We must therefore concern ourselves with the so-called primitive-equation models which do allow the unrestricted propagation and possible amplification of gravity waves. Lacking a fundamental understanding of the real processes which limit gravity oscillations in the atmosphere, we must provide an <u>artificial damping scheme</u> suitable for smothering the unwanted waves and restoring the quasi-geostrophic balance of the model flow.

Many ingenious damping schemes have been proposed in the literature and found to reduce significantly the level of "meteorological noise" in the models. Taking the broader view, however, one may say that no such scheme could provide a satisfactory answer to the problem of data assimilation, for one basic dynamical reason. Artificial damping is needed in the normal course of numerical modelling to replace the small-scale dissipation processes which must occur in the real atmosphere, but which lie beyond the range of scales explicitly represented in the model flow. In other words, the artificial damping must somehow stand for sub-gridscale processes which are, by necessity, not included in the finite number of degrees of freedom of the model flow. For dynamical consistency, the dissipation rate introduced by artificial damping must be commensurate with the transfer rate between wave numbers which would develop naturally within the spectral range of motions explicitly represented in the model. These spectral transfers are determined by the laws of geophysical fluid dynamics and cannot be artificially increased without causing severe discrepancies with regard to the real atmosphere dynamics. The consequence is an unsurmountable dilemma: the rate of artificial damping can be chosen either small enough to match the real atmosphere sub-gridscale dissipation rate or large enough to reduce effectively the meteorological noise - both desirable conditions cannot be met simultaneously.

Increasing the artificial damping rate is inevitably detrimental to the ability of the forecasting model to simulate the real atmosphere and results in serious deviations from reality in data sparse regions where an accurate simulation is needed. On the other hand, decreasing the artificial damping rate results in excessive meteorological noise induced by new data inputs. A consequence of this situation is the a priori unexpected, but a posteriori obvious fact that adding more data into a well-tuned meteorological analysis routine may result in degraded forecasting skill, even though the new data do include additional information which could help determine the real atmospheric initial state. The most spectacular instance of this situation was afforded by the protracted argument regarding the degree of usefulness of satellite temperature profile data. Faced with the difficulty of introducing suitable modifications in their analyses schemes, certain operational forecasting services were led to develop a "Don't bother us with new data" attitude when they found that satellite temperature data inserted in their then current procedures would not improve and could even degrade short and medium-term predictions.

We must conclude then that the provision of artificial damping is not the proper way of reducing the amplitude of the high-frequency oscillations induced by the insertion of heterogeneous, and possibly noisy, data in the course of numerical modelling of atmospheric circulation. The only proper solution is thus to avoid triggering such oscillations by appropriate data conditioning <u>before insertion</u>. This is the topic of the next section on normal mode initialization procedures.

4. <u>NORMAL MODE INITIALIZATION</u>

This method is based on the obvious idea of separating from the outset the contribution of the data to updating the balanced quasi-geostrophic motion of the model flow, from the contribution to the excitation of spurious gravity oscillations, by means of a suitable modal decomposition of the raw correction field resulting from the input of a new data set. Such an approach, based upon normal mode solutions to a linearized version of the forecasting equations, was first used by Flattery (1970) in his original analysis scheme featuring Hough functions which are the solutions of the linearized shallow water equations on a sphere; these functions were first introduced in tidal research. Dickinson and Williamson (1972) later proposed a general method to find the normal modes for an arbitrary general circulation model and a corresponding linear mode filtering scheme for initializing numerical forecasts.

In the linear filtering procedure, the observational data are expanded in terms of the complete set of normal modes for the linearized governing equations, and then the amplitudes of the unwanted computational and gravity modes are set equal to zero. This method does reduce the unrealistic large high-frequency oscillations which occur during the initial period of the forecasts when no a priori filtering is used. But the technique suffers from a deficiency which is apparent from equation (5) above: because of the presence of a non-linear coupling term, the unwanted high-frequency oscillations can be regenerated fairly rapidly in the course of the time integration. This deficiency may be eliminated to a large extent by the so-called <u>non-linear normal mode initialization</u>, in which the time derivatives of the gravity modes coefficients, rather than the initial values of these coefficients, are set equal to zero. Gravity mode coefficients are modified so that the linear tendency corresponding to the diagonal part, $-i\Lambda X$, of the evolution matrix compensates for the contribution $N(X,X)$ from the non-linear interactions between all modes. Details can be found in Machenhauer (1977) and Baer (1977), who first proposed this refinement of the normal mode filtering technique.

This idea is equivalent to the concept developed by Leith (1980), of projecting the initial, noisy, meteorological fields, produced by straightforward interpolation of the observational data, onto the <u>slow manifold</u> constituted by all slowly varying dynamical solutions of the model equations, while excluding all rapidly oscillating states; the latter are represented in configuration space by fast ocillations about the slow manifold.

Various versions of this refined filtering procedure, with or without precautions for minimizing the rejection of original data inputs, have been discussed by Daley and Puri (1980) and a procedure of this kind has been implemented with encouraging results by the European Centre for Medium Range Weather Forecasts. It remains to be seen whether one standard modal expansion, based on the normal mode solutions of the linearized equations about one simple state of the modal flow, i.e., the state of rest, is adequate for all initial meteorological conditions, or whether a much more cumbersome diagnonalization procedure for each individual initial state is actually needed.

A very positive indication of the success of this method for filtering out the meteorological noise from the outset is seen in the remarkable ability of the ECMWF forecasting procedure to accept large amounts of unconventional non-synoptic satellite data and produce a definite improvement over forecasts based on synoptic data only. In this instance, we find at last that more information, even if it is noisy information, yields a more accurate description of the real atmosphere. Nonlinear normal mode initialization is discussed in this volume by R. Daley. Other aspects of the initialization problem are discussed by K. Bube and M. Ghil.

5. THE MATHEMATICAL PROBLEM OF 4-DIMENSIONAL ASSIMILATION

The mathematical basis for understanding the continuous or discontinuous adjustment process involved in 4-dimensional data assimilation is not well established as yet. From our introductory remarks, it appears that one reason for the unavailability of a sound theoretical background may be the lack of satisfactory existence, uniqueness and smoothness theorems for the solutions of the equations of motion in the case of a fully developed turbulent flow. The fact is that forcing successive corrections onto the ongoing time integration of a global numerical prediction model is a purely numerical process which has no counterpart in the physical world. Consequently, the eventual convergence of this process or its divergence cannot be judged on the basis of qualitative physical arguments or limited numerical experimentation. A thorough theoretical approach is a very difficult problem, however. A linear theory of the convergence of one particular dynamical initialization method based on a repeated forcing of the same set of field values on the computed fields obtained in successive forward and backward time integrations, has been attempted by Talagrand (1980). This iterative procedure lends itself readily to a proper definition of convergence, while convincing numerical experiments have shown it to be essentially equivalent to successive forcing of an ongoing forecast with the same amount of external data as provided in the forward-backward assimilation process during a given time interval. This linear theory proved to be inconclusive, however.

A more promising approach has also been tried by Talagrand (1980) when the flow can be legitimately approximated by solutions of the linearized shallow water equations (Eqs. 1-3) or linear combinations thereof; a similar reasoning can also be developed in the case of a multilevel model. This approach is based on the consideration of the perturbation energy invariant:

$$E = \frac{1}{2} \int \left[gh^2 + HV^2 \right] ds \qquad (6)$$

where $\frac{1}{2} gh^2$ is the availabe potential energy of the perturbation and $\frac{1}{2} HV^2$ is its kinetic energy per unit area (assuming an incompressible fluid of unit density). Now, since a linear combination of two solutions of the linearized problem is also a solution, it follows that the quadratic difference:

$$D = \frac{1}{2} \int \left[g(h-h_0)^2 + H (\vec{V} - \vec{V}_0)^2 \right] ds \qquad (7)$$

is also invariant in the course of an unperturbed time evolution of the initial value problems starting with the reference field $(h_0, \vec{V}_0)$, on the one hand, and with the "first-guess" field or trial field $(h, \vec{V})$, on the other. But D is obviously changed by forcing "observational" data taken from the reference model state onto the evolution of the trial field. Indeed, substituting values of parameters h_0 or $\vec{V}_0$ from the reference field at time t, for the corresponding parameters h or $\vec{V}$ in the trial field nullifies the contribution to expression (7) at this time and place. Thus, the straightforward substitution of any subset of reference field values for the corresponding subset of parameters in the trial field always reduces the quadratic difference D between the two model states. The resulting monotone decrease of the quadratic difference does not, by itself, guarantee that the difference will eventually decrease to zero, i.e., that the successive forcing process would converge. It can be shown in the case of backward-forward integration that the process does converge if and only if the "observational data" available during the time interval of one cycle specify a unique solution of the governing equations. This is obviously a necessary condition for convergence; it is also sufficient in this simple case, <u>regardless of the nature of the reference state</u>, viz., whether it is geostrophically balanced or not.

The convergence of 4-dimensional data assimilation towards one particular solution of the model equations does not depend here upon geostrophic adjustment nor damping gravity waves generated by the straightforward substitution of heterogeneous field values. One must indeed consider that the very special role played by the quasi-geostrophic modes in meteorological forecasting is a contingent and indeed very fortunate circumstance which reduce the 4-dimensional assimilation problem to the consideration of the "slow-manifold" instead of the full range of all possible solutions of the equations of geophysical fluid dynamics. One should not forget, however, that this special circumstance is not central to the mathematical problem of the convergence of the numerical procedure of 4-dimensional data analysis.

A somewhat different approach to the problem of 4-D data assimilation, which does address at the same time the preferred nature of slow solutions to the forecast equations, is presented in this volume by M. Ghil and collaborators. The results of applying the particular schemes of 4-D data assimilation and of initialization used at ECMWF to the most complete meteorological data set in existence today are discussed by M. Kanamitsu.

REFERENCES

Baer, F. 1977 Adjustment of initial conditions required to suppress gravity oscillations in non-linear flows. Contrib. Atmos. Phys. 50, 350-366.

Daley, R. and Puri, K. 1980 Four-dimensional data assimilation and the slow manifold. Mon. Weather Rev., 108, 85-99.

Dickinson, R.E. and Williamson, D.L. 1972 Free oscillations of a discrete stratified fluid with application to numerical weather prediction. J. Atmos. Sci., 29, 623-640.

Eliassen, A. 1954 Provisional report on calculation of spatial covariance and autocorrection of the pressure field. Videnskaps Akademiets Institutt for Vaer-og Klimaforskning Report No.5. (reprinted in this volume)

Flattery, T.W. 1970 Spectral models for global analysis and forecasting. Proc. Sixth AWS Tech. Exchange Conf., U.S. Naval Academy, Air Weather Service Tech. Rept. 242.

Gandin 1963 Objective analysis of meteorological fields, Leningrad, Translated from Russian. Israel Program for Scientific Translation, London, Oldbourne Press, 1965.

Leith, C.E. 1980 Non-linear normal mode initialization and quasi-geostrophic theory. J. Atm. Sci. 37, 958-968.

Machenhauer, B. 1977 On the dynamics of gravity oscillations in a shallow-water model, with applications to normal mode initialization. Contrib. Atmos. Phys., 50, 253-271.

Richardson, L.F. 1922 Weather prediction by numerical process. Cambridge Univ. Press, London (Reprinted by Dover, New York).

Schlatter 1975 Some experiments with a multivariate statistical objective analysis scheme. Mon. Weather Rev., 103, 246-257.

Talagrand 1980 On the mathematics of data assimilation. (submitted to Tellus).

A REVIEW OF METHODS FOR OBJECTIVE ANALYSIS
Nils Gustafsson
Swedish Meteorological and Hydrological Institute
Norrköping, Sweden

1. INTRODUCTION

For a given situation, our knowledge about the state of the atmosphere is given by a large number of actual and recent observations, irregularly distributed in space and time. The procedure of combining these observed data to make conclusions about the total variation of the meteorological variables within the area of interest, has generally been called meteorological analysis. The term "objective analysis" has been used for analysis as a numerical procedure by the aid of a computer to distinguish from the manual or "subjective" analysis procedure. In my opinion, the term "objective analysis" is too pretentious, a more relevant terminology is "numerical analysis" versus "manual analysis".

The first generation of numerical schemes, introduced about 30 years ago, generally consisted of simple procedures for two-dimensional interpolation of the observed data into a regular network of gridpoints. To improve the quality of the numerical analyses, a short-range numerical forecast generally was used as a preliminary field ("first guess"). These two-dimensional interpolation procedures were satisfactory for construction of initial analyses for the relatively simple numerical forecast models which were utilized at that time, and since only radiosonde- and surface observations were available, the observing systems did not introduce any further complication. Since then, however, the forecast models have been refined and have also become more sensitive to the quality of the initial analyses. Further, today we have a more mixed and complex observational network with very variable quality of data from various data sources. The introduction of non-synoptic data is another complication. This progress in numerical modelling and observational techniques has necessitated significant development work in the field of numerical analysis methods. Today also numerical forecast models play a significant role in the analysis of observed data, and the term "4-dimensional data assimilation" has been introduced for this complicated procedure.

This lecture will cover one part of what we today call 4-dimensional data assimilation, namely the procedure of transfer of information from the irregularly distributed observations into a form which is convenient for insertion of the information into numerical forecast models and we will use the term "numerical analysis" for this procedure.

The subject of this lecture is mainly restricted to numerical analysis methods for creation of initial data fields to be used in numerical forecast models. It should be mentioned, however, that methods for numerical analysis of meteorological data have a wide application in many other fields, e.g. quality control of climatological data, planning of observational networks and climatological studies. In this context it should be pointed out that the user requirements should be thoroughly considered when designing a certain numerical analysis scheme. A very detailed numerical analysis, e.g. useful for manual short-range weather forecasts, would probably not be the best analysis for initialization of large-scale numerical forecast models.

2. SOURCES OF INFORMATION FOR NUMERICAL ANALYSIS OF METEOROLOGICAL FIELDS

2.1 Observational data

The main sources of information for the numerical analysis of meteorological fields are large numbers of observations. When designing a scheme for numerical analysis the following characteristics of the various types of observations should be considered:

A) Observations are irregularly distributed in space and time.
B) The various types of observations represent different scales of atmospheric variability depending on the different degree of averaging inherent in the various observing systems.
C) Different physical parameters are measured by different observing systems.
D) The accuracy of the various types of observations is quite variable. All measurements are associated with small random and/or systematic errors and also large errors occur in most types of observations.

In table 2.1.1 I have tentatively summarized some of our present knowledge of the characteristics of our operational observing systems. The table is an updated version of a similar table from Bengtsson (1975).

Table 2.1.1: Characteristics of data from operational meteorological observing systems

Observing system	Type of observation	Parameters	Vertical resolution	Horisontal resolution	Time resolution	Random errors	Other errors, Remarks
Surface stations (SYNOP, SHIP)	Synoptic	p_s T_s v_s r_s	– – – –	– – – –	3 hours 3 hours 3 hours 3 hours	±0.1 mb ±1°C ±1 m/s ±5%	Quality of data from commercial ships is variable.
Upper-air soundings (TEMP, PILOT)	Synoptic	T v r	~50 mb ~50 mb ~50 mb	– – –	6-12 hours 6-12 hours 6-12 hours	±1° ±1 m/s ±5%	Larger errors ±2° in the stratosphere Larger wind errors occur e.g. for small radar elevation angles. Errors are vertically correlated.
Aircraft reports (AIREP, ASDAR)	Non-synoptic	T v	– –	500-1000km (AIREP) 150km (ASDAR)	cont.	±3 m/s	Frequent large errors occur in AIREP reports Quality of ASDAR reports is excellent.
Satellite vertical temperature soundings (TOVS)	Non-synoptic	T	~200 mb	500 km (GTS-data)	cont. (6 hours)	±1.5-2°	Larger errors in cloudy areas when only micro-wave measurements are utilized. Errors are horizontally and vertically correlated.
Satellite Sea-Surface temperatures	Non-synoptic	T_s	–	2-10 km	cont.	±1-2°	Systematic errors seem to occur, especially in the tropics (according to ECMWF)
Satellite cloud wind vectors	Non-synoptic	v_{low} (~850mb) v_{high} (~200mb)	– –	500 km 500 km	6 hours 6 hours	±6 m/s ±12m/s	Large systematic errors occur in assignment of heights to the wind vectors.

2.2 Meteorological/physical knowledge and past experience

For the construction of numerical analyses physical/meteorological knowledge and past experience should be utilized in order to construct the best possible numerical analyses. This knowledge and past experience could include:

A) Statistical information on the past behaviour of the atmosphere in the form of, for example, climatological mean values and standard-deviations of occurred deviations from these climatological means. Experience on atmospheric scales could be summarized in atmospheric power spectra or, equivalently, spatial auto-correlations.

B) Knowledge of relationships between meteorological parameters and physical constraints on the variability of meteorological parameters which, to some degree of approximation, are fulfilled in the real atmosphere. Examples of such relationships and constraints are:
   - Geostrophic or gradient wind balance. The non-linear balance equation.
   - Hydrostatic balance.
   - Vertical temperature profiles are only permitted to be weakly superadiabatic.
   - Humidity fields are not permitted to be supersaturated.

C) Experience of atmospheric disturbances summarized, for example, in structure models. An example of such a structure model is the typical distribution of wind, temperature, humidity and clouds in the area of a developing midlatitude cyclone. Experience of this kind has so far been utilized only to a very small extent in numerical analysis schemes.

D) Numerical forecast models with certain abilities to predict the time evolutions of atmospheric phenomena. In order to make full use of the numerical models, past experience with the forecast models could be used and this experience could be summarized in, for example, statistics of past numerical forecast errors.

E) Knowledge of the characteristics of the various observing systems.

F) Research and development in the fields of mathematics, statistics and numerical methods.

3. SUBPROCESSES IN SCHEMES FOR NUMERICAL ANALYSIS OF METEOROLOGICAL FIELDS

Considering the aims of numerical analysis schemes and the various sources of information which were discussed in the previous section, a number of necessary subprocesses in schemes for numerical analysis can be identified:

A) Quality control of the observed data to be utilized. Observations with large errors must be identified, and if possible corrected or eliminated.
B) Filtering of small amplitude random or systematic errors.
C) Space interpolation to a regular network of gridpoints or, in the case where a spectral analysis scheme is utilized, integration of the representing space functions over the irregularly distributed observations.
D) Interpolation in time.
E) Forced adjustment of the meteorological variables by the aid of relationships between the different variables in order to utilize cross-correlated observed information.
{F) Initialization of the analyzed parameter fields before insertion of the data into numerical forecast models. Normally this procedure is not considered as being a part of the numerical analysis scheme.}

When meteorologists construct manual analyses, these subprocesses are performed simultaneously in an intuitive fashion.

In numerical analysis schemes also some of these subprocesses could be performed simultaneously. For example, in a multivariate statistical analysis scheme, the space interpolation, the filtering of small amplitude errors and the forced adjustment between massfield and wind field are performed by one single procedure.

4. REVIEW OF THE DEVELOPMENT OF NUMERICAL ANALYSIS METHODS

In his book about his historical first attempt to compute a numerical weather forecast, Richardson (1922) suggested that the problem of initial data should be solved by using data from one observing station in each of the gridpoints for the forecast calculations. Also when the first experiments with the barotropic forecast model were performed in the late 1940's, numerical analysis was avoided by digitizing of manual analyses. It soon turned out, however, to be necessary to speed up this tedious manual procedure.

Within a few years numerical analysis schemes were independently developed by several groups of meteorologists. Panofsky (1949) and also Gilchrist and Cressman (1954) introduced the polynomial interpolation method. Bergthorsson and Döös (1955) developed the successive correction method and a similar method was also introduced by Cressman (1959). For many years successive correction methods were used in most operational systems for numerical analysis of meteorological fields. This early dominance of the successive correction method has today been taken over by statistical (or optimal) interpolation schemes. Statistical interpolation methods were introduced by Eliassen (1954) and Gandin (1963). Over the past 30 years of development of numerical analysis schemes, most presented schemes belong to one of the following five classes of numerical analysis schemes (some analysis schemes may utilize elements belonging to several of the enumerated classes):

(1) (Local) polynomial interpolation methods in which mathematical functions are locally, in the neighbourhood of the individual gridpoints, adjusted to fit the observed data.
(2) Statistical interpolation methods.
(3) Successive correction methods.
(4) Variational numerical analysis methods introduced by Sasaki (1958).
(5) Spectral analysis methods in which mathematical functions are globally (or hemispherically) adjusted to fit observed data. An operational spectral analysis scheme has been developed by Flattery (1970).

In the following sections I will outline the basic principles of these five types of numerical analysis methods. During my second lecture I will discuss in more detail a number of operational numerical analysis schemes, containing elements of methods from classes (2)-(4).

4.1 Local polynomial interpolation methods

The basic principle of the polynomial interpolation methods is to adjust polynomial functions to the observed data in the close vicinity of the gridpoint, for which analyzed values are required. As an example we will consider two-dimensional analysis of the geopotential with the aid of geopotential and wind observations.

The variation of the geopotential in the vicinity of the gridpoint is approximated by the following polynomial function

$$z(x,y) = \sum_{ij} a_{ij} x^i y^j \quad i,j \geq 0 \quad i+j \leq n \quad (4.1.1)$$

where x and y are e.g. coordinates in a polarstereographic projection. If n=3 we are using cubic surfaces, if n=2 we are using quadratic surfaces.

In order to determine the coefficients a_{ij} of the polynomial function z(x,y) we utilize the least square method to minimize the deviations between the observed data and the corresponding polynomial function values in the vicinity of the gridpoint. In this fitting procedure also wind observations are utilized to adjust gradients of z(x,y).

$$\text{Minimize } E = \sum_{k=1}^{m_z} p_k (z_k^{OBS} - z(x_k, y_k))^2 +$$

$$\sum_{\ell=1}^{m_w} q_\ell \{(u_\ell^{OBS} - u_g(x_\ell, y_\ell))^2 + (v_\ell^{OBS} - v_g(x_\ell, y_\ell))^2\} \quad (4.1.2)$$

where

m_z = the number of selected geopotential observations z_k^{OBS} in the vicinity of the gridpoint

m_w = the number of selected wind observations u_ℓ^{OBS}, v_ℓ^{OBS} in the vicinity of the gridpoint.

p_k and q_ℓ are empirical weighting factors which may be functions of observed parameter, distance between the observation and the gridpoint, data density in the vicinity of the observation point, data accuracy etc. The main purposes of the weighting factors are to make the polynomial fit best in the close vicinity of the gridpoint and to avoid unrealistic behaviour of the polynomial function due to uneven distribution of observations and observational errors.

$u_g(x,y)$ and $v_g(x,y)$ are derived from the geopotential function z(x,y) by the geostrophic relationship:

$$u_g(x,y) = -\frac{g}{f} \frac{\partial z(x,y)}{\partial y}$$

$$v_g(x,y) = \frac{g}{f} \frac{\partial z(x,y)}{\partial x} \quad (4.1.3)$$

The minimizing procedure (4.1.2) will result in a system of linear equations for determination of the coefficients a_{00}, a_{10}, a_{01}, a_{20}, a_{11}, ..., a_{0n}. Note that at least 3 measurements (m_z+2m_w) are required for linear surfaces, 6 measurements are required for quadratic surfaces and 10 measurements are required for cubic surfaces. In order to get any filtering effect on the small amplitude observational errors, however, it is necessary to utilize more pieces of information than the minimum required.

The analyzed value at the gridpoint (x_g, y_g) is obtained by the value $z(x_g, y_g)$ of the derived function $z(x,y)$.

The polynomial interpolation method, exemplified here by two-dimensional interpolation of geopotential, may be generalized to three or four dimensions. In the three-dimensional case, temperature observations can be utilized in the minimization procedure by the aid of the hydrostatic equation and it is possible to obtain vertically consistent analyses. In the four-dimensional generalization, it is possible to utilize also non-synoptic data. In operational schemes based on the polynomial method numerical forecast values also generally enter into the minimization procedure.

To conclude, the polynomial method is non-linear in the sense that non-linear functions are used to approximate the variation of the analyzed variable. It is important to note, however, that the resulting analyzed value in the gridpoint is a linear function of the utilized information and in this respect polynomial interpolation is similar to the other local analysis methods which will be described in following sections.

4.2 Statistical interpolation methods

In the polynomial interpolation method we utilized mathematical functions to approximate the variation of the meteorological variables in the vicinity of the gridpoint. The selection of interpolating functions is quite arbitrary and it is difficult to see how past experience on e.g. atmospheric scales enter into the analysis computations.

In the statistical interpolation methods, past experience about the behaviour of the atmosphere is used as the main source of information for determination of the interpolation weights. Before turning to a discussion about the statistical interpolation methods, we need to introduce some statistical concepts.

4.2.1 Basic statistical concepts

Let f_{it} represent the true numerical value of the meteorological variable f in the point i at the time t. We will consider the true value as the value we obtain after removal of the scales of motion we do not wish to analyze. We will consider these small scale variations as being part of the observational error Δf_{it}. The observed value f_{it}^{OBS} will then be given by

$$f_{it}^{OBS} = f_{it} + \Delta f_{it} \qquad (4.2.1.1)$$

In all statistical treatment the mean-value is of great importance. In order to have any significance a mean-value cannot be too general, the situations over which the mean-value is computed must have something in common.

When describing the statistical structure of the atmosphere we generally use an ensemble mean-value, which means a mean-value over a large number of similar situations. Wellknown mean-values are the monthly or seasonal mean-values, computed from a large number of periods (climatological normal values).

To denote a mean-value in time we introduce the bar-operator ($\overline{A}$):

$$\overline{f}_i = \overline{f_{it}} = \frac{1}{T} \int_T f_{it} \, dt \qquad (4.2.1.2)$$

Here T denotes the time-period (or time-periods) over which the mean-value is computed. In the following we will use the mean-value operator without specifying how the mean-value is computed. For the anomalies, which are the deviations of the true values from the mean-values, we will use the following notation:

$$f'_{it} = f_{it} - \overline{f}_i \qquad (4.2.1.3)$$

Of course it is not sufficient to describe only the mean conditions of the atmosphere. In order to describe the amplitude of the variations of a meteorological variable around its mean-value we use the variance, which is the mean-value of the squares of the anomalies:

$$VAR_i = \overline{(f'_{it})^2} = \overline{(f_{it} - \overline{f}_i)^2} \qquad (4.2.1.4)$$

The square-root of the variance is called the standard-deviation:

$$SD_i = \sqrt{VAR_i} = \sqrt{\overline{(f_{it} - \bar{f}_i)^2}}$$

For use in numerical analysis of meteorological fields a characteristic of the spatial variation of different meteorological variables is of greatest interest. When we are working with one single meteorological variable we use the co-variance or the auto-co-variance to characterize the spatial variations. The co-variance of a variable f between two points i and j is the mean-value of the products of the anomalies in the two points:

$$m_{ij} = \overline{f'_{it} f'_{jt}} = \overline{(f_{it} - \bar{f}_i)(f_{jt} - \bar{f}_j)} \qquad (4.2.1.6)$$

When the two points i and j coincide the covariance reduces to the variance

$$VAR_i = m_{ii} = \overline{(f_{it} - \bar{f}_i)^2} \qquad (4.2.1.7)$$

When we are working with different meteorological variables we use the cross-co-variance:

$$M_{f_i g_j} = \overline{f'_{it} g'_{jt}} = \overline{(f_{it} - \bar{f}_i)(g_{jt} - \bar{g}_j)} \qquad (4.2.1.8)$$

The value of the covariance is to a high degree dependent on the amplitude of the variations forming the covariance. In most cases it is of greater interest to have a relative measure of the spatial variation. Then it is more practical to use the auto-correlation:

$$\mu_{ij} = \frac{m_{ij}}{SD_i \, SD_j} = \frac{m_{ij}}{\sqrt{m_{ii} m_{jj}}} = \frac{\overline{(f_{it} - \bar{f}_i)(f_{jt} - \bar{f}_j)}}{\sqrt{\overline{(f_{it} - \bar{f}_i)^2}\,\overline{(f_{jt} - \bar{f}_j)^2}}} \qquad (4.2.1.5)$$

An auto-correlation value close to +1 indicates a very similar variation of the meteorological variable at the two points.

The auto-correlation can also be generalized to the cross-correlation:

$$\mu_{f_i g_j} = \frac{\overline{(f_{it} - \bar{f}_i)(g_{jt} - \bar{g}_j)}}{\sqrt{\overline{(f_{it} - \bar{f}_i)^2}\,\overline{(g_{jt} - \bar{g}_j)^2}}} \qquad (4.2.1.10)$$

In the concepts introduced so far we have considered deviations of true values from climatological mean values. It is possible, however, to use the same concepts for true deviations from any other preliminary field values, e.g. values obtained by a numerical forecast model. In the case of preliminary field values obtained by a numerical forecast model we will have covariances and correlations for the numerical forecast errors.

$$m_{ij}^{NP} = \overline{(f_{it} - f_{it}^{NP})(f_{jt} - f_{jt}^{NP})}$$

$$\mu_{ij}^{NP} = \frac{m_{ij}^{NP}}{\sqrt{m_{ii}^{NP} \cdot m_{jj}^{NP}}}$$

(4.2.1.11)

In the following discussion of statistical interpolation methods, m_{ij} and μ_{ij} will denote covariances and autocorrelations for the true deviations from any type of preliminary field values f_{it}^{P}:

$$m_{ij} = \overline{(f_{it} - f_{it}^{P})(f_{jt} - f_{jt}^{P})}$$

$$\mu_{ij} = \frac{m_{ij}}{\sqrt{m_{ii} \, m_{jj}}}$$

(4.2.1.12)

4.2.2 Basic statistical interpolation theory

Assume that we want to utilize n observations f_{it}^{OBS} (i=1, ..., n) in the vicinity of a gridpoint g to compute the analyzed value f_{gt}^{NA} in the gridpoint at a certain time t. In statistical interpolation schemes, f_{gt}^{NA} is computed as a linear combination of a preliminary field value f_{gt}^{P} at the gridpoint and the observed deviations $f_{it}^{OBS} - f_{it}^{P}$ from the preliminary field:

$$f_{gt}^{NA} = f_{gt}^{P} + \sum_{i=1}^{n} p_i (f_{it}^{OBS} - f_{it}^{P})$$

(4.2.2.1)

The interpolation weights p_i (i=1, ..., n) are obtained by requiring that the mean square error of interpolation is minimum:

$$\text{Minimize } E = \overline{(f_{gt} - f_{gt}^{NA})^2}$$

(4.2.2.2)

Insertion of (4.2.2.1) into (4.2.2.2) and evaluation of the square will give:

$$E = \overline{(f'_{gt})^2} + \sum_{i=1}^{n} \sum_{j=1}^{n} \{\overline{f'_{it} f'_{jt}} + \overline{\Delta f_{it} f'_{jt}} + \overline{f'_{it} \Delta f_{jt}} +$$

$$+ \overline{\Delta f_{it} \Delta f_{jt}}\} p_i p_j - 2 \sum_{i=1}^{n} \{\overline{f'_{gt} f'_{it}} + \overline{f'_{gt} \Delta f_{it}}\} p_i \quad (4.2.2.3)$$

where $f'_{it} = f_{it} - f^P_{it}$ denotes the deviations of the true values from the preliminary field values or, in other words, the errors of the preliminary field values. $\Delta f_{it} = f^{OBS}_{it} - f_{it}$ denotes the observational errors, including such small-scale variations that we do not want to analyze.

Among the terms in the expression for E, terms of the type $\overline{\Delta f_{it} f'_{jt}}$ denote the cross-covariance between preliminary field errors and observational errors. For most observing systems it is certainly correct to assume that there is no dependence between the observational errors and the preliminary field errors. This will simplify the expression for E since

$$\overline{f'_{gt} \Delta f_{it}} = \overline{\Delta f_{it} f'_{jt}} = \overline{f'_{it} \Delta f_{jt}} = 0 \quad (4.2.2.4)$$

If we further introduce $d_{ij} = \overline{\Delta f_{it} \cdot \Delta f_{jt}}$ for the covariance of the observational errors we will have

$$E = m_{gg} + \sum_{i=1}^{n} \sum_{j=1}^{n} \{m_{ij} + d_{ij}\} p_i p_j - 2 \sum_{i=1}^{n} m_{gi} p_i \quad (4.2.2.5)$$

Necessary conditions for a minimum of E with respect to the interpolation weights p_k (k=1, ..., n) are

$$\frac{\partial E}{\partial p_k} = 2 \{\sum_{i=1}^{n} (m_{ik} + d_{ik}) p_i - m_{kg}\} = 0 \quad (4.2.2.6)$$

$$k=1, ..., n$$

or

$$\sum_{i=1}^{n} \{m_{ik} + d_{ik}\} p_i = m_{kg} \quad (4.2.2.7)$$

$$k=1, ..., n$$

If we know the covariances m_{ik} and m_{kg} of the preliminary field errors and the covariances d_{ik} of the observational errors it is possible to solve the system of n linear equations (4.2.2.7) to obtain the interpolation weights p_k (k=1, ..., n) and by expression (4.2.2.1) it is then possible to compute the analysis f_{gt}^{NA}.

By multiplying equation (4.2.2.7) with p_k and summing for k=1, ..., n we will obtain

$$\sum_{k=1}^{n} \sum_{i=1}^{n} \{m_{ik} + d_{ik}\} p_i p_k - \sum_{k=1}^{n} m_{kg} p_k = 0 \qquad (4.2.2.8)$$

By subtracting (4.2.2.8) from (4.2.2.5) we will obtain the minimized interpolation error

$$E^{MIN} = m_{gg} - \sum_{i=1}^{n} m_{ig} p_i \qquad (4.2.2.9)$$

In practical applications of statistical interpolation it is often convenient to work with normalized deviations from the preliminary field. If we use the standard-deviations $\sqrt{m_{ii}}$ of preliminary field errors as normalizing factors our interpolation formula will read

$$\frac{f_{gt}^{NA} - f_{gt}^{P}}{\sqrt{m_{gg}}} = \sum_{i=1}^{n} p_i \frac{f_{it}^{OBS} - f_{it}^{P}}{\sqrt{m_{ii}}} \qquad (4.2.2.10)$$

$$\sum_{i=1}^{n} \{\mu_{ij} + \gamma_{ij}\} p_i = \mu_{jg} \qquad (4.2.2.11)$$

$$j=1, ..., n$$

$$E^{MIN} = m_{gg}(1- \sum_{i=1}^{n} \mu_{ig} p_i) \qquad (4.2.2.12)$$

where

$\gamma_{ij} = \dfrac{d_{ij}}{\sqrt{m_{ii} m_{jj}}}$ is the normalized covariance of the observational errors.

Any practical application of the statistical interpolation method must include the following elements:

(A) Computation of a preliminary field f^P_{gt} at all gridpoints and all observational points.

(B) Modelling of the preliminary field error covariances $m_{ij} = \overline{(f_{it} - f^P_{it})(f_{jt} - f^P_{jt})}$

(C) Modelling of the observational error covariances:
$d_{ij} = \overline{(f^{OBS}_{it} - f_{it})(f^{OBS}_{jt} - f_{jt})}$

(D) Selection of observed data f^{OBS}_{it} to influence the gridpoint value f^{NA}_{gt}.

(E) Solution of the system of linear equations (4.2.2.11) to obtain the interpolation weights p_k (k=1, ..., n).

In a later part of my lecture I will further discuss these important elements of any statistical interpolation scheme. For the moment I will restrict myself to mention the various types of experience and knowledge that it is possible to include into statistical interpolation schemes:

(1) The variability and the scale of the atmospheric variable to be analyzed is described by the preliminary field error covariance m_{ij}.

(2) The characteristics of the observational errors is given by observational error covariance d_{ij}.

(3) It is possible to utilize any relationship between the different meteorological variables in the construction of the preliminary field error covariances.

Further, it should be pointed out that the geometrical configuration of the observations around the gridpoints is automatically taken care of by the covariance matrix and that the preliminary field value is given a weight in accordance with its accuracy.
It is also necessary to stress that, by statistical interpolation, the interpolation error is minimized over a large number of cases and that this does not mean that the interpolation weights are the "best" for each individual case.

4.2.3 Properties of covariance and correlations

The covariance of a meteorological field f is a function of the position vectors r_i and r_j (e.g. $r_i = (x_i, y_i, p_i)$ in the p-system) of the two points for which the covariance is valid:

$$m_{ij} = m(r_i, r_j)$$

We shall now define the properties **statistical homogeneity** and **statistical isotropy** of a meteorological field in reference to the covariance m_{ij}. A meteorological field is **homogeneous** in reference to the covariance if the covariance is independent of a translation of the two positions. This means that the covariance is only depending on the difference between the positional vectors $m(r_i, r_j) = m(r_i - r_j)$. Homogeneity in reference to the covariance implies that the variance is constant since $m(r_i, r_i) \equiv m(0)$.

A meteorological field is **isotropic** in reference to the covariance if the covariance is independent of a rotation in the field around the centre point on the line between the two positions, that is $m(r_i, r_j) = m((r_i + r_j)/2, |r_i - r_j|)$.

Homogeneity and isotropy of a meteorological field in reference to the covariance implies that the covariance is dependent on the distance between the points only. For the auto-correlation homogeneity and isotropy can be defined in a similar way.

4.2.4 Examples of atmospheric statistical structures.

For any practical application of the statistical interpolation method it is necessary to construct a model for the spatial covariances to be used in the scheme. To simplify the construction of these covariances models it is an advantage if the homogeneity and isotropy conditions are fulfilled.

From long series of observations the covariances and the correlations of different meteorological variables can be computed and from these empirical covariances and correlations it is possible to construct mathematical models which fit the empirical data with a certain degree of approximation.

We will now study some examples of computed statistical structures and we will limit ourselves to the study of correlations. Besides the spatial variations of the variances the covariances express the same thing as the correlations. On a global scale the variations of variances are large, e.g. the maximum of the variance of the 500 mb geopotential is about $(150 \text{ m})^2$ while the variance in the tropics is only about $(20 \text{ m})^2$. Since the correlation is normalized for the variations in the variances its properties are "smoother" and the conditions of homogeneity and isotropy are better fulfilled.

Figure 4.2.4.1 illustrates computed auto-correlations for the 500-mb-geopotential between a reference station (Hannover) and other radio-sonde stations in Europe and the Atlantic. From the figure we can conclude that isotropy is well fulfilled for distances below 3000 km in reference to the correlation and it is possible to use a correlation model where the correlation is a function of distance between the observations only.

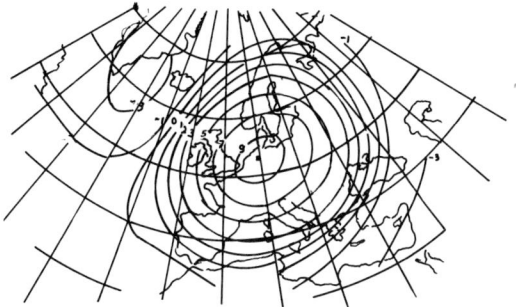

Figure 4.2.4.1.: Isolines for the auto-correlations of the 500 mb
geopotential between the station in Hannover and
surrounding stations.
From Bertoni and Lund (1963)

For the use in three-dimensional analysis systems it is also possible to compute three-dimensional auto-correlations or cross-correlations. Figure 4.2.4.2. shows isolines of computed cross-correlations between the 500 mb geopotential in station 01 384 (marked R in the figure) and the surface pressure in surrounding stations. The translation of the center of isotropy illustrates the mean tilt of atmospheric circulation systems with the surface low, in the mean, situated to the south-east of the upper trough.

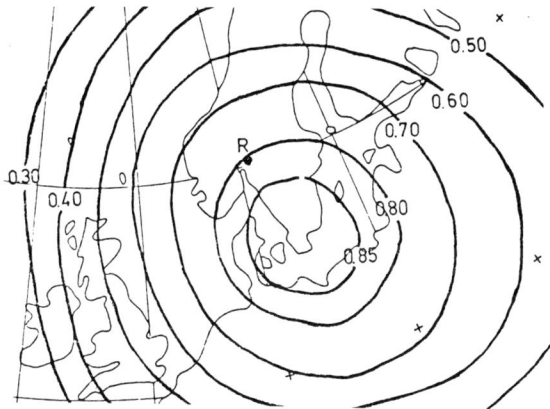

Figure 4.2.4.2.: Isolines of the cross-correlation between the 500 mb
geopotential in station 01 384 (R) and the surface
pressure in surrounding stations.

The autocorrelations for the temperature have similar properties as the auto-correlation for the geopotential.

The auto-correlation for the wind has a more complex structure. In figure 4.2.4.3. isolines for the auto-correlation of the u-component (dashed line) and isolines for the auto-correlation of the v-component (full line) are plotted. The isolines are stretched in the direction of the component.

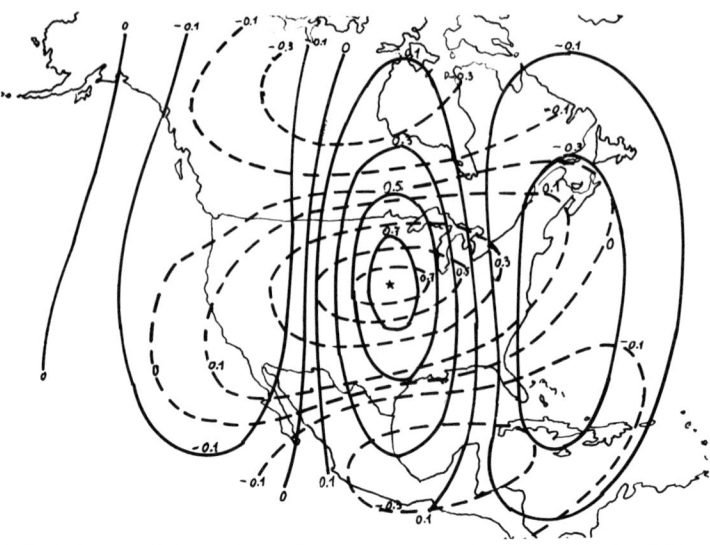

Figure 4.2.4.3.: Isolines for the auto-correlation of the 500 mb u-wind component (dashed line) and the auto-correlation of the 500 mb v-wind component (full line). The "star" indicates the position of the reference station. (From Buel (1972).

It is difficult to construct a correlation model directly from the observed correlations based on wind data. It is possible, however, to derive correlation models for wind components which approximate fairly well the observed correlations by application of the geostrophic relation to isotropic correlation models for the geopotential, e.g.:

$$\overline{u'_{it} u'_{jt}} = \frac{g^2}{f_i f_j} \overline{\frac{\partial z'_{it}}{\partial y_i} \frac{\partial z'_{jt}}{\partial y_j}} = \frac{g^2}{f_i f_j} \frac{\partial^2 \overline{z'_{it} z'_{jt}}}{\partial y_i \partial y_j} = \frac{g^2}{f_i f_j} \frac{\partial^2 m(r_{ij})}{\partial y_i \partial y_j}$$

where m (r_{ij}) is the isotropic covariance for the geopotential which is a function of distance r_{ij} only.

Figure 4.2.4.4 illustrates schematically the behaviour of the functions for auto-correlations and cross-correlations derived by the geostrophic wind equation from an isotropic correlation function for the geopotential. In multivariate statistical interpolation schemes, these types of correlation functions are generally utilized.

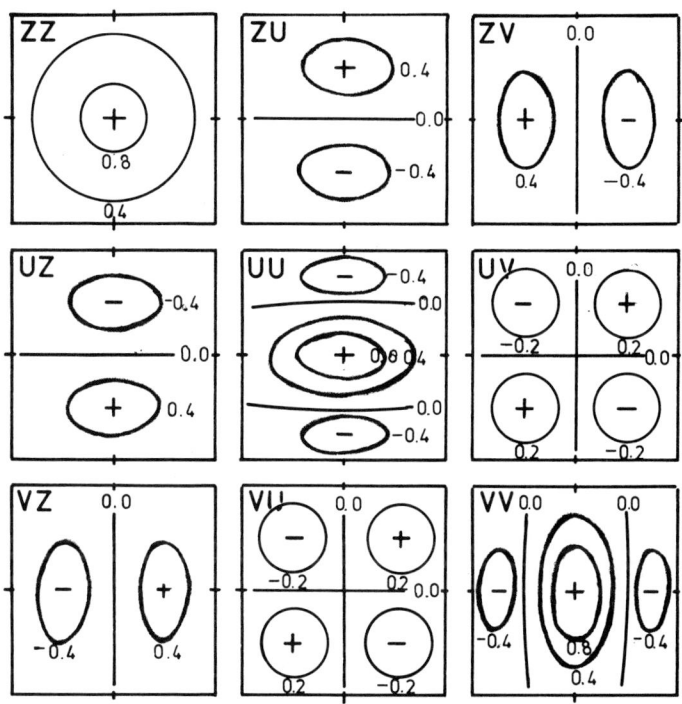

Figure 4.2.4.4.: Schematic illustration of correlation functions and cross-correlation functions for multi-variate analysis derived by the geostrophic assumption.

It needs to be mentioned that the scales of motion in the atmosphere, expressed by e.g. auto-correlation functions, have a latitudinal dependence and also, especially on the boundaries of the tropics, a seasonal variation.

This fact is illustrated in Figure 4.2.4.5 where (isotropic) autocorrelation functions for the u-wind component during winter and summer at 25^0N have been plotted.

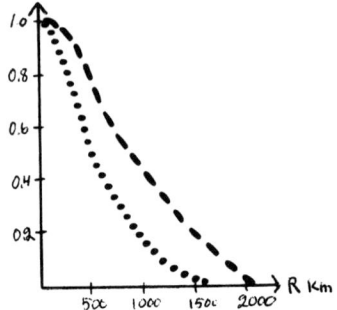

Figure 4.2.4.5: Isotropic auto-correlation functions obtained from u-wind component at 25^0N according to Alaka (1972) (.......=summer---=winter)

Finally, Figure 4.2.4.6 illustrates the auto-correlation of daily precipitation values in Arjeplog (marked A in the figure) and surrounding stations. The isolines are stretched in the direction of the Scandinavian chain.

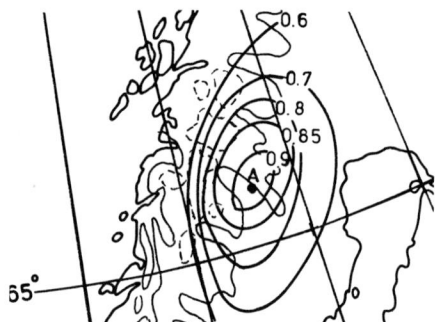

Figure 4.2.4.6: Isolines for the auto-correlation of daily precipitation values in Arjeplog, Sweden (marked A) and surrounding stations. The seasons are winter and spring.

For statistical interpolation schemes, based on numerical forecast fields as preliminary fields, it is necessary to obtain covariance models for the numerical forecast errors. This problem will be discussed in a following chapter.

4.3. Successive correction methods

The first step in the successive correction method is to construct a preliminary field f_{gt}^P for all the gridpoints to be analyzed. The preliminary field could, e.g. be computed as a weighted mean of a climatology field $\bar{f}_g$ and a valid numerical forecast field f_{gt}^{NP}.

The basic idea in the successive correction method is to correct this preliminary field iteratively during several analysis "scans". In each analysis scan the corrections are formally computed in the same way as in the statistical interpolation method, thus by interpolation of the deviations of the observed values from the preliminary field:

$$f_{gt}^{NA} = f_{gt}^P + \sum_{i=1}^{n} \alpha_i (f_{it}^{OBS} - f_{it}^P) \qquad (4.3.1)$$

In the correction method the interpolation weights are computed explicitely without any solving of a system of linear equations. In the previous analysis system of the Swedish Weather Service, the successive correction interpolation weights were given by formulas of the following form:

$$\alpha_i = \frac{\mu(r_{ig}) \cdot h(\rho_i)}{\alpha_p + \sum_{i=1}^{n} \mu(r_{ig}) \cdot h(\rho_i)} \qquad (4.3.2)$$

The most important factor of the weight is the distance-dependent function $\mu(r_{ig})$ where r_{ig} is the distance between the gridpoint and the observational point. The $\mu(r)$ -functions are very similar to the auto-correlation functions. Since the observations may be very unevenly distributed around the gridpoint it is necessary to correct the weights for observations which are highly correlated. This is taken care of by the station-density function $h(\rho_i)$ where ρ_i is the number of influencing observations within a certain distance from observation i. Finally the weights are normalized by the sum of the weights. In order to give some weight to the preliminary field in data-sparse areas a term α_p is also added to the normalizing factor.

One application of (4.3.1) for all gridpoints results in an analyzed field which in turn can be treated as a preliminary field (or a

"second guess" field) and the whole procedure can be repeated. By
a number of such correction scans it is possible to have a successive
adjustment of the gridpoint-values to the observations.

By using different distance-dependent functions, $\mu(r)$, it is also
possible to correct for errors of different atmospheric scales
during the different scans and the number of scans is determined by
the range of atmospheric scales of the variable to be analyzed. In
figure 4.3.1. the $\mu(r)$-functions which were used for the analysis
of the surface pressure and the 500-mb geopotential at the weather
service in Sweden are plotted.

The successive correction method is rather empirical in nature. In
the statistical interpolation method all the empirical correction
factors, which are used in the successive correction method, are
taken care of in an automatic and optimum way. In this sense the
successive correction method may be looked upon as an empirical
approximation to the statistical interpolation method. The main
advantage of the successive correction method is the relatively low
number of computations compared with the statistical interpolation
method.

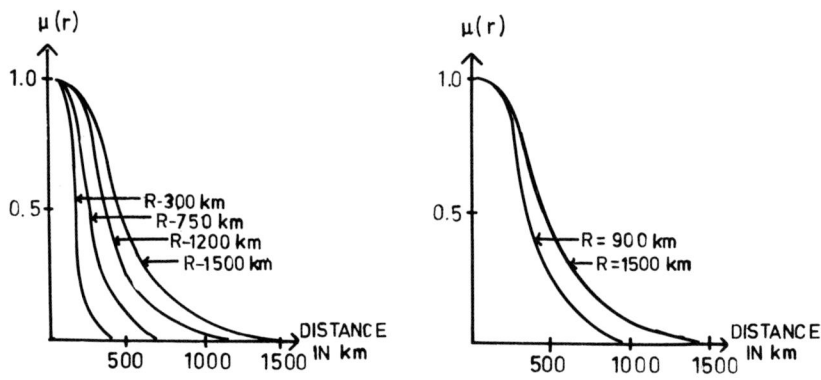

Figure 4.3.1: Weighting functions $\mu(r)$ for the numerical analysis
of surface pressure (left) and 500 mb geopotential
(right). In each scan stations within R km are allowed
to influence the analyzed values in the grid-points.

4.4 Comparisons of the performance of the polynomial, statistical and successive correction interpolation methods.

There are certain difficulties in comparing the performance of different space interpolation methods. In my opinion, the best way of comparison is to run numerical forecast models with initial analyses produced by the different numerical analysis schemes to be compared, and to evaluate the relative quality of the resulting numerical forecasts ("the proof of the pudding lies in the eating"). Ideally, this is the normal procedure of testing new operational analysis systems before the replacement of old systems.

A much simpler method is to interpolate from observations in a group of stations to a station which was not utilized for the interpolation, and to compare the interpolated values with the corresponding observed values in this station. Gandin (1963) has reported about such a comparison between the polynomial, the statistical and the successive correction interpolation methods. In areas of dense data network all methods gave very close results. For sparse data network (7 stations, the mean distance between the stations being about 1300 km) the result is contained in table 4.4.1. The empirical root-mean square error is given as well as a theoretical error estimated by the aid of a covariance function and the variance of the observational errors. The results in table 4.4.1 indicates that the statistical interpolation method gave the best result for the sparse data network and that polynomial interpolation gave the worst result. It is characteristic that the 2nd degree polynomial gave a worse result than the 1st degree polynomial. Higher degree polynomials are more sensitive to irregular station distributions and to random observational errors.

Table 4.4.1: The root-mean-square errors of comparison between the interpolated values of the heights at 500 mb on the basis of observational data from 7 stations and the observed ones for the sparse network of stations.

Method of interpolation	Errors of comparison (gpm)	
	Empirical	Theoretical
Statistical interpolation	50	51
Correction method (distance weighting functions)	64	63
1st degree polynomial	72	71
2nd degree polynomial	153	116

These factors may create unrealistic behaviour of the polynomials between the observational points. It should be added, however, that operational schemes based on the successive correction method or the polynomial method, contain empirical correction functions which were not used in Gandin's comparison.

4.5 Variational numerical analysis methods

The basic idea of the variational numerical analysis methods was proposed by Sasaki (1958) and further developed by him for e.g. the analysis of the structure of tropical cyclones.

In order to minimize the errors of the analyzed fields of different meteorological variables they may be mutually adjusted to each other. By adjustment we mean that some given relationship between the variables is forced to be, partly or completely, fulfilled. Sasaki proposed that calculus of variations should be used to optimize this adjustment procedure.

Assume that we have analyzed the fields of the basic meteorological variables by some space interpolation method: geopotential (Z), temperature (T), wind components (U and V) and humidity (Q). Between these variables some given relationships can be assumed to be valid:

$$F_i(Z,U,V,T,Q) = 0 \quad (4.5.1)$$
$$i = 1,\ldots,n$$

It may be possible to arrive at a complete or approximate fulfilment of these relationships by adding small corrections (dZ, dU, dV, dT, dQ) to the original analyses (Z^{NA}, U^{NA}, V^{NA}, T^{NA}, Q^{NA}). The magnitudes of the corrections are made as small as possible by minimizing the analysis volume integral of the square of these corrections:

$$\text{Minimize } I = \iiint_{x\,y\,p} \{\alpha_z(dZ)^2 + \alpha_u(dU)^2 + \alpha_v(dV)^2 +$$

$$\alpha_T(dT)^2 + \alpha_Q(dQ)^2\} \, dx \, dy \, dp \quad (4.5.2)$$

under the constraints

$$F_i(Z^{NA} + dZ, U^{NA} + dU, V^{NA} + dV, T + dT, Q + dQ) \approx 0 \quad (4.5.3)$$
$$i=1, \ldots, n$$

The optimization problem (4.5.2) + (4.5.3) is a typical problem to be solved by calculus of variations. The methods of solving the optimization problem may be formulated to have the constraints completely fulfilled (strong constraints) or only approximately fulfilled (weak constraints).

In a complete system for numerical analysis of meteorological fields, it is necessary to use the variational analysis technique together with some space interpolation method. It is an advantage if the coefficients α_z, α_u, α_v, α_T, α_Q in the minimizing integral could be functions of the quality of the corresponding analyses, e.g. by using the mean square interpolation errors obtained by a statistical interpolation method.

In a following chapter a simple application of the variational analysis technique used in the analysis system of the Swedish Weather Service will be described.

Variational techniques also have a wide application for initialization of numerical forecast models.

4.6 Spectral analysis methods.

In spectral analysis methods, the actual state of the atmosphere is represented by an expansion in a series of space-dependent functions with time-dependent coefficients. The analysis procedure consists of the determination of those time-dependent coefficients which makes the series expansion best fit the observed data. Analysis methods of this type are used operationally by NMC in Washington, see Flattery (1970) and Hayden (1976), and by the Meteorological office in UK, see Dixon (1976). In the analysis system of the UK Met. Office, the state of the atmosphere is expanded into series of global orthogonal polynomials. As an example of the spectral analysis technique, we will shortly describe the NMC spectral analysis method.

If λ, ϕ and p are longitudes, latitude and pressure respectively the global representation of geopotential and wind components is given by

$$\begin{cases} z(\lambda,\phi,p) = \sum_{\ell=0}^{24} \sum_{m=1}^{24} \sum_{n=1}^{7} \left[a_{\ell mn} \cos(\ell\lambda) + b_{\ell mn} \sin(\ell\lambda) \right] H_{\ell m}(\phi) E_n(p) \\ \\ u(\lambda,\phi,p) = \sum_{\ell=0}^{24} \sum_{m=1}^{24} \sum_{n=1}^{7} \left[a_{\ell mn} \cos(\ell\lambda) + b_{\ell mn} \sin(\ell\lambda) \right] U_{\ell m}(\phi) E_n(p) \\ \\ v(\lambda,\phi,p) = \sum_{\ell=0}^{24} \sum_{m=1}^{24} \sum_{n=1}^{7} \left[a_{\ell mn} \cos(\ell\lambda) - b_{\ell mn} \sin(\ell\lambda) \right] V_{\ell m}(\phi) E_n(p) \end{cases}$$

(4.6.1)

The horizontal functions are eigen-functions of the linearized shallow water equation on a rotating sphere (Hough-functions) and the vertical functions are empirical orthogonal functions, which are eigen-functions of the vertical covariance matrix. By using only 7 vertical functions, some of the small-scale vertical variations are filtered.

The height analysis is performed simultaneously with the wind analysis in order to relate height gradients to the wind field. The height-functions are related to the wind-functions by near geostropic balance.

The coefficients $a_{\ell mn}$, $b_{\ell mn}$ are obtained by simultaneous minimization of integrals of the form

$$I_z = \int_V \{z^{OBS} - Z(\lambda,\phi,p)\}^2 \, dv$$

$$I_u = \int_V \{u^{OBS} - u(\lambda,\phi,p)\}^2 \, dv \quad (4.6.2)$$

$$I_v = \int_V \{v^{OBS} - v(\lambda,\phi,p)\}^2 \, dv$$

with a weighting

$$\zeta \cdot I_z + I_u + I_v \quad (4.6.3)$$

where V denotes the total volume of the atmosphere from the surface up to 50 mb and ζ denotes the relative weight of height observations versus wind observation in the minimization procedure.

We will obtain the minimum value by application of the standard least-square technique

$$\frac{\partial}{\partial a_{\ell mn}} \{\zeta \cdot I_z + I_u + I_v\} = 0$$

$$\frac{\partial}{\partial b_{\ell mn}} \{\zeta \cdot I_z + I_u + I_v = 0 \quad (4.6.4)$$

$$\ell=1, \ldots, 24; \; m=1, \ldots, 24; \; n=1, \ldots, 7$$

I will not go into details about how this minimization procedure is further evaluated, but I will just describe and discuss how the observed data enters into the procedure. The solution of (4.6.4) will include computations of integrals of the form

$$T^c_{\ell mn} = \int_V \{\zeta Z^c_{\ell mn} Z^{OBS} + U^c_{\ell mn} U^{OBS} + V^c_{\ell mn} V^{OBS}\} dV \qquad (4.6.5)$$

where

$$Z^c_{\ell mn} = \cos(\ell\lambda) \, H_{\ell m}(\phi) E_n(p)$$
$$U^c_{\ell mn} = \cos(\ell\lambda) \, U_{\ell m}(\phi) E_n(p) \qquad (4.6.6)$$
$$V^c_{\ell mn} = \cos(\ell\lambda) \, V_{\ell m}(\phi) E_n(p)$$

In order to compute the integral $T^c_{\ell mn}$, the total global analysis volume is divided into a number of boxes, and in each box an average of the observed values is computed to participate in the computation of integral (4.6.5).

The procedure of computing this "average observed value" in each integration box is similar to what is performed in conventional local grid-point analysis schemes. Thus, we do not avoid the local averaging of reports or local "interpolation" by applying spectral analysis techniques.

In the NMC spectral analysis system the computation of the average observed value in each integration box is a purely empirical weighting, based on numerical experiments. The net effect of the weighting gives less weight to satellite or aircraft data when radiosonde data are available in the same integration box, and reduces the weight of observations of any type when the data density is large. This weighting system assumes that satellite sounding data have larger error levels than do radiosonde data, but the weighting does not discriminate between random errors and errors which have a spatial correlation.

The procedure is applied iteratively with a 12 hours numerical forecast as a first guess in the first iteration. During each iteration, "analyzed" values from the previous iteration will dominate the computation of integrals (4.6.5) in data sparse areas. 9 iterations are needed to obtain a reasonable adjustment of the analyzed values to the observations.

In the operational application the integral (4.6.3) is minimized twice, once with a value of ξ appropriate for the height analysis and once with a value appropriate for the wind analysis. This procedure produces two sets of coefficients; one for the heights, the other for winds. Numerical values of ξ are chosen so that heights receive more weight than winds in the height analyses, winds more weight than heights in the wind analyses.

5. CRITICAL REVIEW AND COMPARISON OF OPERATIONAL SCHEMES FOR NUMERICAL ANALYSIS OF METEOROLOGICAL FIELDS

In the previous chapters the basic principles of numerical analysis were introduced and some general methods for numerical analysis were presented and discussed. In this chapter I will make a critical review and comparison of some operational systems for numerical analysis of meteorological fields. In table 5.1 I have tried to summarize some characteristics of the operational systems at some selected weather services together with clear plans for replacements of these systems by new systems. Due to the delay in exchange of information between weather services, the table may not be correct in all its details.

It can be noted from table 5.1 that many weather services today have introduced statistical interpolation schemes and that other weather services are in the progress to replace existing analysis schemes with statistical interpolation schemes. In the following I will first highlight some general problems in the formulation of statistical interpolation schemes and then, in the light of this discussion, I will review and compare the operational systems of Canada, USA and Sweden.

5.1 Operational statistical interpolation schemes

5.1.1 Basic problems in statistical interpolation schemes

Formally, all operational statistical schemes are subsets of the following general 3-dimensional multi-variate statistical interpolation scheme:

$$A_{gt}^{NA} = A_{gt}^{P} + \sum_{i=1}^{n} p_i (A_{it}^{OBS} - A_{it}^{P}) \qquad (5.1.1.1)$$

where the symbol A denotes any of the basic meteorological variables (pressure, height, temperature, wind components or humidity) or variables derived from these basic variables (e g layer thickness or layer mean temperature). Thus, A_{gt}^{NA} denotes the analyzed value, at a time t, of any of these variables in a gridpoint g of the three-dimensional network of gridpoints covering the analysis volume and A_{it}^{OBS} any observed values in the 3-dimensional vicinity of this gridpoint. A^{P} is the preliminary field and in most operational systems it consists of a short-range numerical forecast field.

Table 5.1 Characteristics of operational numerical analysis schemes.

Organization or country	Present operational analysis methods	Analysis area	Analysis forecast cycle	Plans
Australia	. Successive correction method . Variational blending techniques	. S Hemisph . Regional	12 hours 6 hours	
Canada	. Multivariate 3-dimensional statistical interpolation	. N Hemisph . Regional	6 hours (3 hours for the surface)	
France	. Successive correction method, windfield and massfield balance through first guess fields . Multivariate 3-dimensional statistical interpolation	. N Hemisph . Regional	6 hours	
F.R.G.	. Successive correction method. Upper-air analyses are built up, level by level, from the surface . Variational height/wind adjustment	. N Hemisph	12 hours (6 hours for the surface) Climatology only as preliminary fields	Multivariate statistical interpolation is being developed
Japan	. Successive correction method. . Height field analyses are corrected by wind analyses	. N Hemisph . Regional	12 hours	Multivariate statistical interpolation is tested
Sweden	. Uni-variate 3-dimensional statistical interpolation . Variational height/wind adjustment	. N Hemisph . Regional	12 hours 3 hours	A multivariate scheme is being tested
U.K.	. Hemispheric orthogonal polynomial method . Uni-variate statistical interpolation (repeated insertion of data)	. Global	6 hours	Multivariate schemes considered
U.S.A.	. Spectral 3-dimensional analysis . Multivariate 3-dimensional statistical interpolation	. Global . Global	6 hours	
U.S.S.R.	. 2-dimensional statistical interpolation	. N Hemisph	12 hours	
E.C.M.W.F.	. Multivariate 3-dimensional statistical interpolation	. Global	6 hours	

Generally the humidity field is treated separately and analyzed by a univariate interpolation scheme. If we eliminate humidity, we will have the problem of simultaneous mass- and windfield analysis.

Minimization of the mean square interpolation error will result in a system of linear equations for the interpolation weights p_i (i=1, ..., n) and the minimum mean square interpolation error may also be determined:

$$\sum_{i=1}^{n} \{m(A_i,A_j) + d(\Delta A_i, \Delta A_j)\} p_i = m(A_j,A_g) \qquad (5.1.1.2)$$
$$j=1, ..., n$$

$$E^{MIN} = m(A_g,A_g) - \sum_{i=1}^{n} m(A_g,A_i) p_i \qquad (5.1.1.3)$$

where

$$m(A_i,A_j) = \overline{(A_{it}-A_{it}^P)(A_{jt}-A_{jt}^P)}$$ denote the 3-dimensional cross-covariances of the preliminary field errors

and

$$d(\Delta A_i, \Delta A_j) = \overline{(A_{it}^{OBS}-A_{it})(A_{jt}^{OBS}-A_{jt})}$$ denote the 3-dimensional cross-covariances of the observational errors. When deriving (5.1.1.2) we have assumed that observation errors and forecast errors are not cross-correlated.

The design of any statistical interpolation analysis scheme will involve the following decisions and tasks:

(1) Decision on which meteorological variables to be analyzed
(2) Decision on which observed quantitites to enter into the interpolation
(3) Decision on any "splitting" of the analysis procedures to reduce the required computer resources.
(4) Modelling of the preliminary field error covariances
(5) Modelling of the observational error covariances
(6) Development of algorithms for selection of influencing information for each gridpoint value.
(7) Decision on which numerical procedure to be used for the solution of the system of linear equations (5.1.1.2).

In the following we will discuss some of these problems and tasks. The discussion is to some extent based on the "Minutes of a Workshop on Optimum Interpolation" (NMC, Washington, 19-20 September, 1977).

5.1.1.1 Selection of meteorological variables to be analyzed by statistical interpolation schemes

The selection of meteorological variables to be analyzed is in most cases quite obvious, we will select those variables which are required by the actual numerical forecast model. However, it is appropriate to be a little cautious and not to select too model dependent variables. The following considerations should be taken into account:

(1) It is an advantage to have as good direct relationships as possible between the different analyzed variables as well as between the analyzed variables and the observed quantities which enter the multivariate scheme in order to take advantage of large cross-correlations. As an example, it is easier to analyze geopotentials and wind components due to the direct relationship between winds and geopotential gradients rather than to analyze temperatures and wind components between which we only have a vertical differential relationship. Moreover, and this is perhaps more important, the relatively smaller scale of temperature at fixed levels also favours the multivariate analysis of wind and geopotentials.

(2) The possibility for users, other than numerical forecast modellers, to utilize the numerical analyses is simplified if some standard vertical coordinate system, e.g. the p-system, is used for the analysis computations.

(3) The complexity of the statistical structures of the variables to be analyzed should be taken into account. As an example, the statistical structure of meteorological variables on p-surfaces is less complicated than the statistical structure of corresponding variables on σ-surfaces. (In practice this is no problem since statistical structures on σ-surfaces may be obtained from statistical structures on p-surfaces).

As regards selection of a vertical coordinate system for the analysis, the advantages of the p-system should be considered in relation to the possible disadvantages of extra vertical interpolations.

5.1.1.2 Selection of observed quantitites to be utilized in statistical interpolation schemes

Having selected the meteorological variables to be analyzed in a statistical interpolation scheme, it is generally a rather straight forward procedure to select the observed quantities to be used as predictors. There should be a strong relationship between the analyzed variables and the observed quantities or combinations of observed quantities.

I will make just a few remarks:

(1) The combined effect of surface pressure observations and observations of the thickness between the surface and the analysis level is equivalent to the effect of geopotential observations at the analysis level.

(2) Wind observations at one level, e g aircraft observations at ≈200 mb, and a number of adjacent thickness observations between this level and the analysis level will provide useful information on the wind at the analysis level. This is especially true if the errors of thickness observations are horizontally correlated, a characteristic which will make the error of observed thickness gradients (or thermal wind) smaller.

Considerations of this kind should also be taken into account when algorithms for selection of influencing information at the invidual gridpoints are designed.

5.1.1.3 Possible splitting of the fully three-dimensional multivariate statistical interpolation scheme

A satisfactory selection of influencing information for the fully three-dimensional multivariate statistical interpolation scheme will normally result in large covariance matrices which have to be inverted and the available computer resources may not be sufficient. Also other computational restrictions may not allow the implementation of the complete scheme. Thus, for computational reasons, it may be necessary to simplify the analysis scheme.

5.1.1.4 Modelling of the preliminary field error covariances

Most statistical interpolation schemes are based on numerical forecast fields as preliminary fields and it is therefore necessary to obtain the covariances of numerical forecast errors. The structure of the numerical forecast errors at a certain location and at a certain time is a function of several factors:

(1) The uncertainty of the initial data of the forecast
(2) The predictability of the actual state of the atmosphere.
(3) Physical and numerical limitations of the forecast model

In principle, it is possible to include also characteristics of the forecast errors, starting from estimates of the errors in the initial data, into the forecast models and we arrive at the concept of stochastic-dynamic forecast models (Epstein and Pitcher (1973)).

Due to the large computational resources needed, however, nobody has yet tried this approach operationally to obtain covariance of the forecast errors for analysis purposes. Ghil (1980) has reported about development work in this area.

In practice we have to construct models for the forecast error covariances based on historical forecast errors and simplifying assumptions. I will discuss the conventional approach to obtain these models of the covariances.

Generally it is convenient to construct models for the correlations and to treat the standard-deviations separately:

$$m(A_i, A_j) = \sqrt{m(A_i, A_i)} \sqrt{m(A_j A_j)} \; \mu(A_i, A_j)$$

First I will consider the univariate correlations of geopotential forecast errors. Figure 5.1.1.4.1 shows computed correlations of surface pressure forecast errors in the Swedish numerical forecast model. It can be concluded that isotropy is reasonably well fulfilled.

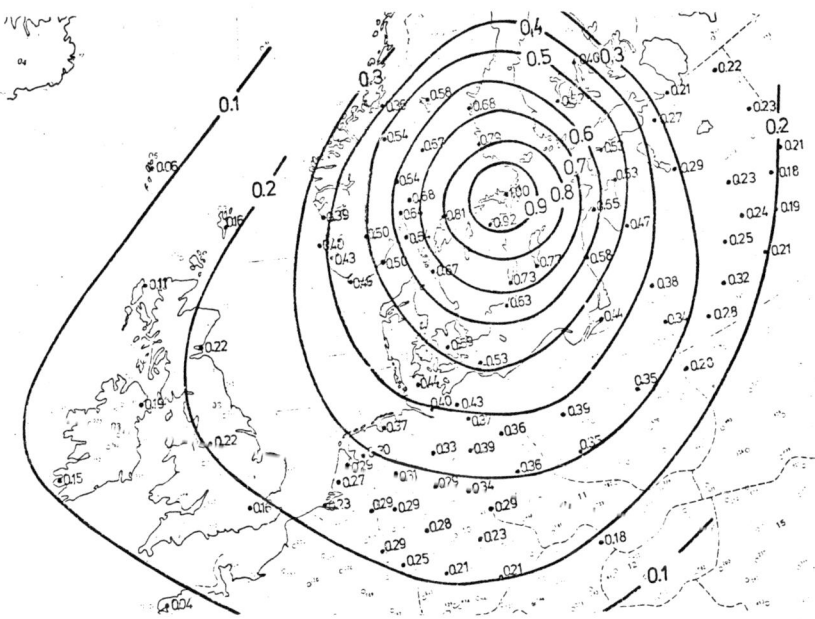

Figure 5.1.1.4.1 Auto-correlation of errors in 12h numerical forecasts of surface pressure in a reference station (Stockholm) and other stations.

Also for the Swedish model, 3-dimensional correlation, e.g. between surface pressure and 500 mb geopotential forecast errors, were computed. The structure of these computed cross-correlations was less regular, but "isotropy" was fulfilled to some extent with the maximum correlation located at the surface below the 500 mb reference station. These empirical correlations may justify a model assumption to separate the fully 3-dimensional correlation into a product of vertical and horizontal correlations. For the geopotentials, the horizontal correlations can be assumed to be functions of distance only

$$\mu(z'(x_i,y_i,p_i), z'(x_j,y_j,p_j)) = \alpha(p_i,p_j) \cdot \beta(r_{ij})$$

where

$r_{ij} = \sqrt{(x_i-x_j)^2 + (y_i-y_j)^2}$ is the horizontal distance between the points (x_i,y_i) and (x_j,y_j).

To my knowledge, all operational 3-dimensional interpolation schemes utilize this model assumption. It was introduced mainly for computational convenience and on the basis of mean characteristics of the numerical forecast errors (see above). The physical meaning is that we assume the forecast errors to have no vertical tilt and in certain situations this assumption will certainly be less than valid. In the case, of for example, fast developing cyclones, which are not predicted by the model and which are described by surface observations only, our analysis model will underestimate the baroclinicity and this may cause considerable forecast errors, especially for detailed short-range forecasts. Bengtsson (1980) has discussed the possibility to utilize observed pressure tendencies to estimate the tilt of developing cyclones.

In most schemes for statistical interpolation the vertical correlations are given by a table of correlations between the analysis levels. Hollett(1975) has computed vertical forecast error correlations, see table 5.1.1.4.1, with special emphasis to eliminate the effect of vertically correlated observational errors.

	850	700	500	400	300	250	200	150	100
850	1.00								
700	.92	1.00							
500	.66	.82	1.00						
400	.51	.66	.93	1.00					
300	.35	.50	.77	.90	1.00				
250	.25	.39	.62	.71	.89	1.00			
200	.13	.24	.42	.45	.63	.88	1.00		
150	.08	.18	.32	.35	.51	.74	.94	1.00	
100	.14	.23	.36	.44	.54	.56	.58	.72	1.00

Figure 5.1.1.4.1 Vertical correlations of 12h geopotential forecast errors according to Hollett (1975).

The conventional procedure to obtain the horizontal correlations $\beta(r)$ is to fit empirical correlations to distance dependent functions. It is necessary to be cautious in the selection of the mathematical functions, especially when correlations for the wind are to be derived from the geopotential correlation functions. The power spectra, corresponding to the correlation function, should preferably be positive for the parameter itself as well as for derivatives of the parameter. For a further discussion on the characteristics of various correlation functions I refer to several papers by Thiebaux (1975, 1976). The most commonly used correlation function is the Gaussian exponential function

$$\beta(r) = e^{-r^2/2b^2} \qquad (5.1.1.4.1)$$

where r is the distance and b^2 a coefficient defining the scale of the parameter to be analyzed.

The forecast errors at the different vertical levels have different scales and therefore we would like to use different horizontal correlation functions at the different levels. In doing so, it is necessary to compute the 3-dimensional correlation in such a way that the resulting covariance matrices are positive definite. Cats (1980) has suggested the following consistent function for 3-dimensional correlations

$$\mu(x_i,y_i,p_i,x_j,y_j,p_j) = e^{-r_i^2/(b_i^2+b_j^2)} \{B_{ij} \frac{2b_i b_j}{b_i^2+b_j^2}\} \qquad (5.1.1.4.2)$$

where b_i^2 and b_j^2 are the coefficients in the correlation functions of (5.1.1.4.1) at the levels p_i and p_j.

B_{ij} is a positive definite n x n matrix corresponding to the analysis levels with unity in the diagonal.

As concerns standard-deviations of the forecast errors, it is also possible to derive these from historical forecast errors, see figure 5.1.1.4.2.

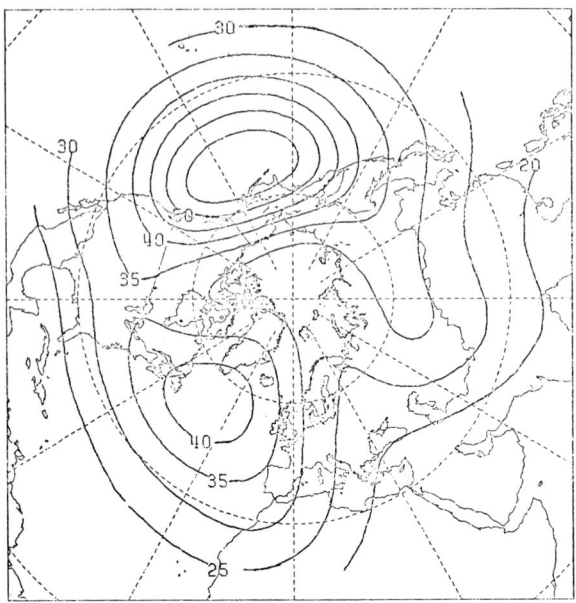

Figure 5.1.1.4.2: Standard deviations of 12 hours 500 mb geopotential forecast errors with the Swedish forecast model. A manual adjustment was applied to arrive at the smooth structure in the figure.

The application of a time-independent standard deviation field is less valid today when we have an observing system which is variable in time. The information amount in, for example, satellite data inserted a few hours ago must be taken into account when new data are inserted. One simple method of "remembering" the amount of recent information was suggested by Bengtsson and Gustafsson (1971) and is today implemented in some operational analysis schemes. Each time the analysis is performed the variance of the analysis error is obtained from the estimated mean interpolation error

$$m_{gg}^{NA}(t) = m_{gg}^{NP}(t) - \sum_{i=1}^{n} p_i \, m_{ig}^{NP}(t) \qquad (5.1.1.4.3)$$

The variance of the forecast error at a certain time $(t+\Delta t)$ is estimated by a linear increase of the forecast error

$$m_{gg}^{NP}(t+\Delta t) = m_{gg}^{NA}(t) + \frac{\Delta t}{T} m_{gg} \qquad (5.1.1.4.4)$$

where m_{gg} is the climatology error variance and T the time-period to reach this error level in the forecast model. To very crudely simulate the advection of forecast errors and observed information, it is possible to apply some smoothing operator to m_{gg}^{NP}.

So far we have only discussed models for the correlations of geopotential forecast errors. For the wind field analysis, a similar technique could be utilized to model observed forecast error correlations. However, as was discussed in a previous chapter, the structure of the wind field (error) covariances is more complex and does not fulfil isotropy, for example. The most common approach to obtain correlations for the wind forecast errors is therefore to apply the geostrophic wind equation to the isotropic correlation functions for the geopotential, e.g:

$$m(u_i, v_j) \cong m(-\frac{g}{f_i} (\frac{\partial z}{\partial y})_i, \frac{g}{f_j} (\frac{\partial z}{\partial x})_j) \qquad (5.1.1.4.5)$$

In the same way, cross-correlations between wind components and geopotential can also be derived. For analysis in tropical areas it is then necessary to correct these cross-correlations for the non-validity of the geostrophic wind equation in the tropics, e.g.

$$m(z_i, u_j) \cong m(z_i, -\frac{g}{f_j} (\frac{\partial z}{\partial y})_j) \cdot \gamma(\text{latitude}) \qquad (5.1.1.4.6)$$

where γ (latitude) describes the fulfilment of the geostrophic wind equation at different latitudes.

Application of (5.1.1.4.5) means that we assume non-divergent winds and that the correlation of the streamfunction is isotropic. Depending on the method for selection of influencing information, the divergent part of the wind will also be analyzed to a certain extent. In order to make a consistent analysis of the divergent part of the wind, however, it is necessary to refine the correlation model by, for example, utilizing the statistical structure of the velocity potential.

5.1.1.5 Modelling of observational error covariances

By comparing observations from any new observing system with adjacent observations from a reference observing system with known error characteristics it is possible, at least in principle, to determine the error characteristics of the new observing systems. Such systematic studies of new observing systems are necessary in order to make full use of the information from the new observing systems. In practice, however, results from such systematic studies are often not available and we have to rely on reasonable "guesses" of what the error characteristics are. Table 2.1 of chapter 2 contain a tentative summary of our present knowledge of the error structures of the various operational observing systems. It should be noted, however, that some of the figures in table 2.1 represent only the measurement errors. If the figures are to be used in statistical interpolation schemes, these figures must be enlarged to cover also the amplitude of the small-scale variations which are not of interest in the analysis. For analysis of synoptic scale surface pressure, as an example, the assumed standard-deviation of errors in surface pressure observations should rather be 1.0 mb than 0.1 mb.

It is of special importance to correctly model the systematic observational errors, e g the horizontally correlated errors of the satellite sounding observations. We can illustrate this importance by a simple example. Bergman (1979) has determined the horizontal correlation of satellite sounding temperature errors to be approximated by the following function

$$\mu(r) = e^{-11.3 \cdot 10^{-6} r^2} \qquad (5.1.1.5.1)$$

where the distance is given in km. In the case where the standard-deviation of the observational error is $2^\circ C$, it is easy to show that the RMS-error of the 100 km horizontal gradient, computed from adjacent satellite temperature reports, will be of the order 0.6°.

Thus, due to the **horizontal** correlation of the observational errors we have a better accuracy in the gradients of the temperatures (the thermal wind) and this information is important when only wind data on a few isolated levels are available. In order to make full use of this information, it is necessary to include correlation models of the type (5.1.1.5.1) in the analysis schemes.

5.1.1.6 Algorithms for selection of influencing information

The quality of the analysis will to a large extent depend on which observed values are selected to influence the analyzed value at the gridpoint. When 2-dimensional univariate schemes were used, it was generally noted that it was sufficient to select 6-8 influencing observations to reach a minimum error level- if more observations were selected, no information was added. For these schemes it was satisfactory to select information from the closest stations, possibly with some extra constraint on the distribution of the selected stations.

The selection of influencing information for the multivariate 3-dimensional schemes is a much more difficult task. Observed quantities which may have very small correlation with the analyzed variable may be of importance due to large cross-correlations with other observed quantities. I have already mentioned a few valuable combinations of observed quantities, other combinations could be added. From the discussion it can be concluded that it is certainly not satisfactory to select information only on the basis of correlations between observed and analyzed quantities.

Two types of selection algorithms are used in present operational analysis systems:

(A) Select, in principle, all observed information in a certain analysis volume, invert the large covariance matrix based on these observations and use the inverted matrix to compute the analyzed values of all variables in the central part of the volume for which observations were selected. ECMWF uses this type of data selection scheme.

(B) Make a local selection of influencing information for each gridpoint, each analysis level and, possibly, each analyzed variable. Utilize empirical algorithms to select also cross-correlated information.

In my opinion the first type of selection algorithms is the most attractive one, provided that the necessary computer resources to invert the large covariance matrices are available.

As concerns local selection of influencing information, it is generally not possible to select more than 8-12 pieces of influencing information due to the cubic increase of computer time as a function of this number. It is a very difficult, if not impossible, task to construct a fast computer algorithm which selects the best possible, or approximately best possible, combination of 8-12 observed values from the enormous number of combinations which are possible with our complex observing systems.

5.1.2 The operational analysis system of the Canadian Meteorological Service

The analysis system of the Canadian Meteorological Service is based on 3-dimensional multivariate statistical interpolation of forecast errors. The system was developed by Rutherford (1975) and it was the first system of this kind that was used operationally.

Analysis cycle: 6 hours
Analyzed variables: U,V,Z and T (multivariate); $T-T_d$ (univariate)
Splitting: All reports are vertically interpolated to create vertically complete reports and then 2-dimensional multivariate interpolation is performed for each analysis level. The vertically interpolated values are influenced by the forecast errors, but this effect is consistently taken care of during the **horizontal** analysis.
Forecast error standard deviations are functions of vertical level only.
Forecast error correlations are modelled as described in section 5.1.1.4.
Observational errors are assumed to be random without any vertical or horizontal correlation.
Data selection: The 20 closest reports are preselected at each level and from all the observed (or interpolated) values in these 20 reports, the 8 values with largest correlation with the gridpoint-value to be analysed are selected for influence.
Critical comment: The splitting is motivated by computational consideration and handled in a consistent way. The selection procedure seems to be too simple to utilize valuable cross-correlations among observed quantities. Furthermore, it is difficult to see how, for example, the splitting handles the useful combined effect of surface pressure observations and satellite thickness observations.

5.1.3 The NMC(USA) operational analysis system

The analysis system of the National Meteorological Centre in USA is based on 3-dimensional multivariate statistical interpolation of forecast errors. The system was developed by Bergman (1979).

Analysis cycle: 6 hours

Analyzed variables: Terrain pressure and tropopause pressure (univariate); U, V, layer-mean temperatures (multivariate); specific humidity (univariate).

No splitting is performed.

Forecast error standard deviations are functions of time, level and position. Linear growth of forecast error in time is assumed.

Forecast error correlations are modelled as described in section 5.1.1.4 utilizing thermal wind equation for the wind/temperature relationship.

Observational errors are modelled so as to have vertical correlations (radiosonde and satellite sounding data) as well as horizontal correlations (satellite sounding data).

Data selection: The 10 values with largest "weight" are selected, the "weight" being a function of both correlation with the gridpoint value and the accuracy of the observed value.

Critical comment: The selection of temperature and wind to enter the multivariate scheme makes the data selection more difficult due to derivated relationship between the variables. The data selection procedure is too simple, valuable cross-correlated reports may not be utilized.

5.1.4 The operational analysis system of the Swedish Weather Service

The analysis system of the Swedish Weather Service is based on univariate 3-dimensional statistical interpolation of forecast errors.

Analysis cycles: 12 hours (N Hemisphere), 3 hours (Limited Area)

Analyzed variables: z, u, v, T and relative humidity.

Splitting: (1) 3-dimensional analysis of mass-field
(2) 3-dimensional analysis of wind-field
(3) 2-dimensional adjustment to gradient wind balance at each analysis level. The estimated interpolation errors are used during this adjustment procedure.

F̱o̱ṟe̱c̱a̱s̱ṯ_e̱ṟṟo̱ṟ_s̱ṯa̱ṉḏa̱ṟḏ-ḏe̱v̱i̱a̱ṯi̱o̱ṉs̱ are functions of level, position and season.

F̱o̱ṟe̱c̱a̱s̱ṯ_e̱ṟṟo̱ṟ_c̱o̱ṟṟe̱ḻa̱ṯi̱o̱ṉs̱ are modelled as described in 5.1.1.4.

O̱ḇs̱e̱ṟv̱a̱ṯi̱o̱ṉa̱ḻ_e̱ṟṟo̱ṟs̱ are in general assumed to be random. For satellite sounding thicknesses a **horizontal** correlation is assumed.

Ḏa̱ṯa̱_s̱e̱ḻe̱c̱ṯi̱o̱ṉ: For the wind analysis, 5 wind vectors (10 observed values) are selected according to the largest correlations. The mass-field analysis is performed 2-dimensionally in data-dense areas. In data-sparse areas, up to 10 observed values are selected and among these 3 surface pressures and 3 satellite thicknesses are selected to utilize the combined effect of these reports.

C̱ṟi̱ṯi̱c̱a̱ḻ_c̱o̱m̱m̱e̱ṉṯ: The variational adjustment sometimes works less satisfactory, especially for small-scale surface pressure analysis. The effect of the small temperature gradient errors in satellite sounding data is not optimally utilized in the wind analysis. A positive effect of the splitting is that it is easier to construct satisfactory selection algorithms.

5.1.4.1 Variational adjustment between wind analyses and geopotential in the operational analysis system of the Swedish Weather Service

In section 4.5 the general principle of variational numerical analysis was presented. As an example, variational adjustment to geostrophic balance between wind- and geopotential analyses is described below. In the operational application at the Swedish Meteorological Service, adjustment to gradient wind balance is performed.

Let Z^{NA} denote the analyzed geopotential field and let U^{NA} and V^{NA} denote the analyzed wind components. In order to obtain geostrophic balance we add adjustment corrections δZ, δU and δV to the analyzed fields:

$$Z = Z^{NA} + \delta Z; \quad U = U^{NA} + \delta U; \quad V = V^{NA} + \delta V \qquad (5.1.4.1.1)$$

We require

$$\hat{U} = \frac{fU}{gm} = -\frac{\partial Z}{\partial y} \quad \text{and} \quad \hat{V} = \frac{fV}{gm} = \frac{\partial Z}{\partial x} \qquad (5.1.4.1.2)$$

where m is the map magnification factor and $\hat{U}$ and $\hat{V}$ are introduced to simplify the derivation of the adjustment equation.

The adjustment equation is determined by minimization of the integrated squares of the adjustment corrections.

Minimize I =

$$\iint_A \{(\delta\hat{U})^2 + (\delta\hat{V})^2 + (\frac{K_z E_{\hat{V}}}{E_z})(\delta z)^2\} dx\, dy \qquad (5.1.4.1.3)$$

where

E_z = mean square interpolation error of the geopotential
$E_{\hat{V}}$ = mean square interpolation error of the wind vector and
K_z an empirical "tuning" coefficient (presently = 0.25).

By utilizing (5.1.4.1.1) and (5.1.4.1.2) we obtain

$$I = \iint_A \{(\frac{\partial z^{NA}}{\partial y} + \frac{\partial \delta z}{\partial y} + \hat{U}^{NA})^2 + (\frac{\partial z^{NA}}{\partial x} + \frac{\partial \delta z}{\partial x} - \hat{V}^{NA})^2 + (\frac{K_z E_{\hat{V}}}{E_z})(\delta z)^2\} dx\, dy$$
$$(5.1.4.1.4)$$

or

$$I = \iint_A F(x, y, \delta z, \frac{\partial \delta z}{\partial x}, \frac{\partial \delta z}{\partial y})\, dx\, dy \qquad (5.1.4.1.5)$$

Minimization of I with lateral boundary conditions $\delta z = 0$ is a general problem in calculus of variations and the solution is given by Euler-Lagrange's equation:

$$\frac{\partial F}{\partial \delta z} - \frac{\partial}{\partial x}\left(\frac{\partial F}{\partial(\frac{\partial \delta z}{\partial x})}\right) - \frac{\partial}{\partial y}\left(\frac{\partial F}{\partial(\frac{\partial \delta z}{\partial y})}\right) = 0 \qquad (5.1.4.1.6)$$

For our special case the solution is given by

$$\nabla^2 \delta z - \{\frac{K_z E_{\hat{V}}}{E_z}\}\delta z = \frac{\partial \hat{V}^{NA}}{\partial x} - \frac{\partial \hat{U}^{NA}}{\partial y} - \nabla^2 z^{NA} \qquad (5.1.4.1.7)$$

With lateral boundary conditions $\delta z = 0$ it is possible to obtain the solution of (5.1.4.1.7) by, **for example, any relaxation procedure.**

It is possible to apply the adjustment procedure described above only to the analysis increments $z^{NA} - z^P$, $U^{NA} - U^P$ and $V^{NA} - V^P$. An application of this type is illustrated in figures 5.1.4.1-3. Preliminary fields for the geopotential and the wind analysis were obtained from 0 hour numerical forecast fields. Note the corrections in the vicinity of Iceland produced during the two analysis phases.

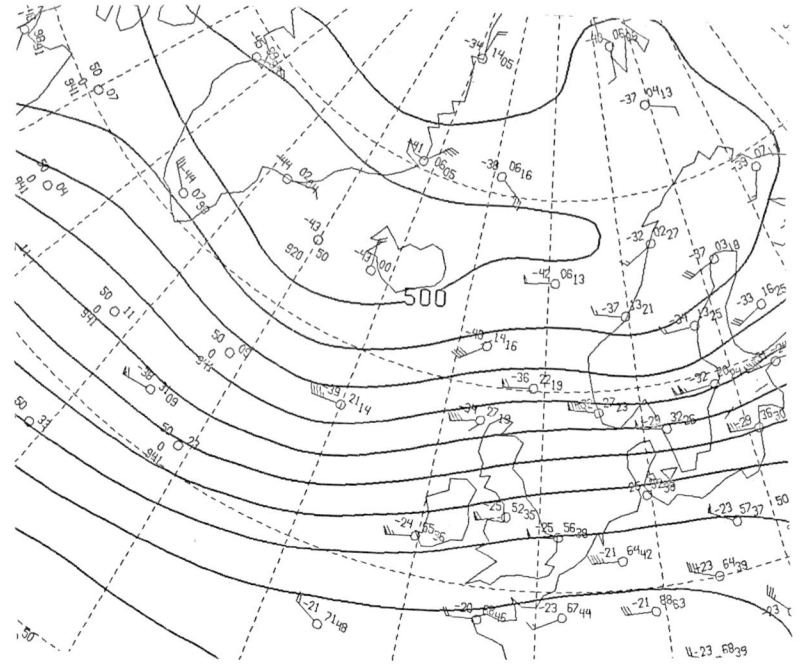

Figure 5.1.4.1: 500mb geopotential numerical forecast for 7 January 1979 18 GMT + 6 hours ("first guess" field)

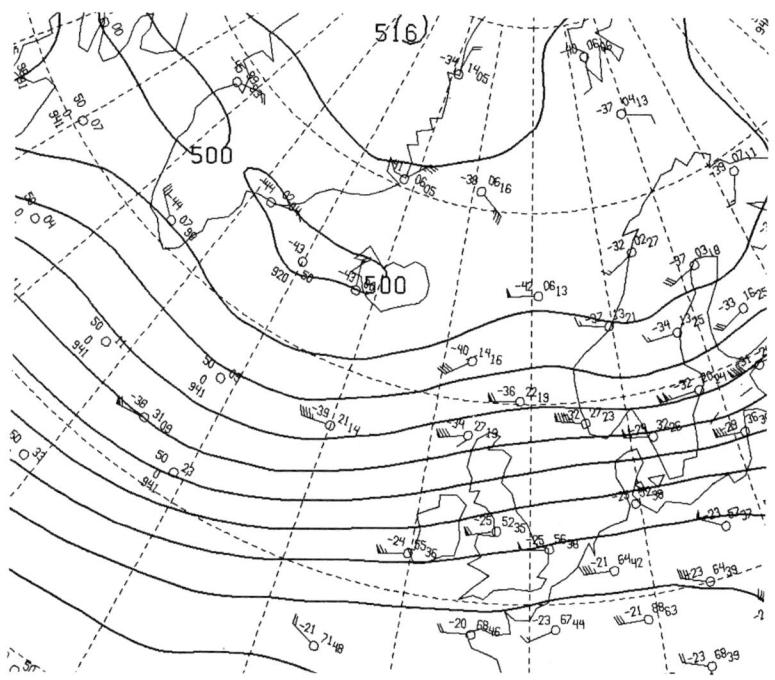

Figure 5.1.4.2: 500mb geopotential analysis for 8 January 1979 00 GMT. Only mass-field analysis was performed.

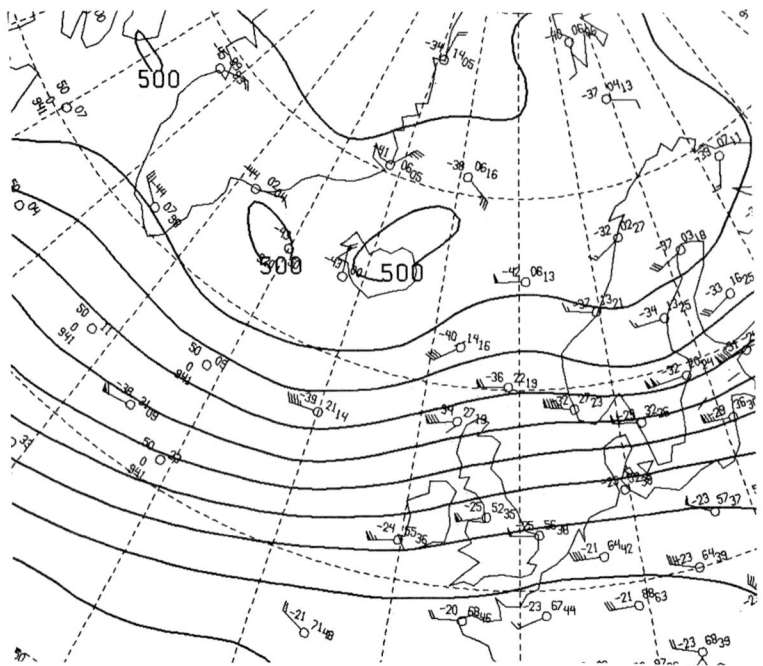

Figure 5.1.4.3: 500mb geopotential analysis for 8 January 1979 00 GMT. Mass-field analysis, wind-field analysis and variational adjustment of analysis increments to geostrophic balance were performed.

6. ANALYSIS OF THE HUMIDITY AND OTHER PARAMETERS OF SMALLER SCALE

6.1 Analysis of the humidity

In the development of numerical analysis schemes, relatively little attention has been paid to the analysis of the humidity field. Possibly this lack of development can be explained by the following factors:

(1) Dynamical simulations have indicated that numerical forecast models are relatively insensitive to the accuracy of the initial humidity fields. During a time period of 12-24 hours, the humidity field adjusts to the vertical and **horizontal** circulation in the model atmosphere.

(2) Radiosonde measurements of humidity are contaminated by small-scale variations that we do not want to analyze.

Today our limited area forecast models are able to produce very detailed weather forecasts, including also precipitation forecasts, for time-periods of 12-36 hours. To get useful precipitation forecasts for these time scales it is important to utilize reasonable initial humidity fields, since the dynamical adjustment process in the model will not have time enough to create the necessary balance. Further, the observations from the dense network of surface stations contain useful information about the humidity of the free atmosphere and in the near future we will also have operational satellite observations of the humidity field. Thus, it is today necessary to pay some more attention to the improvement of our schemes for numerical analysis of the humidity field.

The refined forecast models we are using today are able to develop humidity patterns which are very realistic and these humidity patterns are very difficult to re-construct from the humidity observations only. It can be concluded that the structure of the humidity forecast fields should be preserved during the analysis process, while typical forecast errors like phase speed errors should be corrected with the aid of available humidity observations. Atkins (1974) has described a humidity analysis scheme based on this idea. Atkins used the successive correction method, with a humidity forecast as first guess field and with horizontal weighting functions which were stretched along structures in the forecast humidity pattern. In this way it was possible to preserve e.g. gradients of the forecast humidity field near frontal zones. Atkins also showed improved numerical forecast quality with the aid of these improved humidity analyses.

Kaestner (1974) has developed an analysis scheme in which the humidity analyses are obtained in the following computational steps:

(1) Use geopotential analyses to integrete the ω-equation to obtain diagnostical vertical velocities.

(2) Compute first guess relative humidities from the vertical velocities with the aid of regression-equations.

(3) Obtain estimates of the upper-air relative humidities by regression formulas from parameters in surface reports (cloudiness, cloud types, present and past weather).

(4) Analyse relative humidity by the successive correction method. In data-sparse areas, the humidity values obtained from the surface reports are also utilized.

Most operational schemes for analysis of the humidity field are based on statistical interpolation. In a recent paper (1980), van Maanen discussed the analysis of the humidity field by statistical interpolation methods. I will review his discussion.

First van Maanen discussed the selection of humidity variable to be used for statistical interpolation. The ideal humidity variable has the following characteristics: it has a normal distribution, it is not correlated with the temperature and it is the same variable as used in the forecast model. The normal distribution is an advantage for use in statistical interpolation and the correlation with temperature should be low to prevent the system from effectively making a temperature analysis instead of a humidity analysis. Van Maanen decided to use the mixing ratio for the humidity analysis, mainly since the statistical distribution was found to be close to the normal. To avoid a large correlation with the temperature field, van Maanen used a first guess for the humidity analysis obtained by regression from the temperature. The observed humidity deviations from this first guess field are independent of the temperature field. From a long series of data, van Maanen estimated the horizontal correlations of the humidity with isotropic correlation functions and with the aid of these correlation functions, van Maanen was able to construct humidity analyses in good agreement with the synoptic weather patterns.

6.2 Numerical analysis of the meteorological parameters on the surface weather charts

At all operational weather services a large amount of manual effort is today spent on the slow manual analysis of the surface weather charts. Opinions on the necessity of this resource-consuming manual effort are quite divided among meteorologists. Some meteorologists state that a large part of these resources could be better utilized if they were devoted to making weather forecasts, to interpretation of numerical forecasts and to presentation of the forecasts to the public. Other meteorologists are of the opinion that manual analysis of all the surface charts is necessary in order for the meteorologists to "penetrate the synoptic situation".

Independent of our opinion on this delicate matter, however, it is today at least possible for our numerical analysis systems to produce powerful tools to assist the meteorologists in their manual analysis work. Several weather service (e g Holland and Sweden) are analyzing surface pressure each 3rd hour and by utilizing predicted pressure tendencies together with tendency observations it is also possible to construct consistent tendency analyses. With the aid of other model parameters, e g precipitation, it is also possible to obtain consistent preliminary fields for these parameters. Several years ago, a successful experiment to produce surface charts with pressure, pressure tendency and 3 hours forecasted precipitation amounts was carried out at the Swedish Weather Service. When similar maps were produced operationally, after the 3 hours analysis cycle was put into operation, it turned out, however, that frequent errors occurred in the plotted tendency fields. From a sample of low quality maps it was considered necessary to improve the data checking procedures.

7.   QUALITY CONTROL OF OBSERVATIONAL DATA

Meteorological information that has been observed, coded to meteorological messages, transmitted by the telecommunication links and decoded by some computer program contain errors of different types and of different magnitudes. Some of these errors we may call natural errors since they have been caused by small measurement errors and by small-scale variations of the meteorological variables. These natural errors are always present in the observations and the effects of these errors are reduced by the filtering procedures which are included in all methods for numerical analysis.

However, there are also larger errors which must be detected, corrected or eliminated during the processing of the observations. It has been estimated that about 5-10% of all the surface-based observations contain these larger errors. Most of these large errors are introduced by manual mistakes or during transmission of data over telecommunication lines without transmission error control.

The gradual development of the WWW will hopefully result in a decrease in the number of large errors. The following factors favour such a development:

(1) The introduction of automatic measurements and automatic coding of automatic as well as manual measurements.
(2) The introduction of **computerized** telecommunication systems with possibilities to detect all transmission errors.
(3) The introduction of meteorological quality control of data as close as possible to the original data source (e g by microprocessors at the observing stations) and before the data are inserted into the Global Telecommunication System.
(4) The introduction of new codes for the meteorological messages, e g the new SYNOP-code which will be used operationally from 1 January 1982.

The possibility of checking meteorological observations is based upon the redundancy in the information. A simple example of this redundancy is the reporting of both heights and temperatures in the TEMP messages. In most cases the redundancy is not **complete and then it is** only possible to determine the likelihood of a given observed value. In this **sense** almost all methods for quality control of meteorological observations are based on statistics.

The observations, which are received at a weather service, have been **coded into** meteorological messages. These messages have to be identified, decoded and sorted. This process is called pre-processing or Automatic Data Extraction (ADE). The automatic decoding of the messages is a very **laborious** process since it is necessary to consider the frequent formal errors in the messages like the omission, mixing and truncation of groups of information. The formats of the messages have been agreed upon internationally and some of the formats contain unique control digits. These control digits may be used for simple format control procedures.

In each message there is also a certain degree of redundancy which makes different types of internal checking methods possible. An example of an efficient internal checking method is the hydrostatic control of radiosonde data based upon the redundancy of the temperature and geopotential information in the TEMP messages.

A number of efficient quality control methods are based upon the continuity of the meteorological variables in space and time. During the space continuity control the observed value is generally compared with an interpolated value, based on observations in the vicinity of the observation to be tested. Methods for space continuity control of observational data are discussed in 7.2 below.

It is also possible to utilize dynamical knowledge for the checking of meteorological observations. An example of a dynamical checking method is to check the observed winds against geostrophic winds obtained by interpolation of geopotential observations. Also this dynamical checking methods has a statistical character since the information which is going to be checked as well as the information which is used for the checking of the observations may be in error.

Different quality control procedures will provide us with different types of error indications. Gandin (1971) has introduced the concept of complex quality control for the simultaneous application of several interacting quality control procedures. The complex quality control method increases our possibility to detect and especially correct errors in the observational data.

In some situations it may be very difficult to decide with computerized methods whether an observation is correct or not. Some observations have to be classified as "suspect". In such cases it may be advantageous to use manual intervention in the quality control process. Manual intervention is discussed in section 7.3.

7.1 Quality control of surface observations (SYNOP and SHIP)

The redundancy of the elements reported in the SYNOP-and the SHIP-messages is not as complete as the redundancy in the TEMP-messages. Consequently it is not possible to construct an internal checking method for these messages that detects and corrects errors as efficiently as the hydrostatic control of TEMP-messages.

Further, the percentage of large errors is much higher especially for the SHIP-messages. However, there are a number of checking methods which can be used for SYNOP- and SHIP-reports. Each of these methods may not give too reliable a result if they are applied individually. By utilizing the combination of results from several checking methods the probability of making a correct control decision is greater. Here we will list just a number of possibilities for checking SYNOP- and SHIP-reports.

a) <u>Internal consistency control</u>: A very common error in the surface reports is the omission and mixing of groups of information in the messages. Since there exist no control digits in the message formats it is not possible to detect these errors during the decoding of the reports.

   However, since there are logical or physical relation-ships among the elements in the surface reports, an error of this kind generally results in physical or logical inconsistencies. These errors may thus be detected by different consistency algorithms.

b) <u>Time consistency control</u>: Since the surface observations are reported with high resolution in time a time continuity control of different elements may be very efficient. An efficient time consistency check is to compare successive pressure and pressure tendency observations.

c) <u>Checking of the positions of merchant ships</u>: The ship code contains information of the position of the ship as well as the speed and the course of the ship. This allows for a check of the movements of ships between consecutive reporting hours.

d) <u>Space continuity control of different elements</u>.

e) <u>Checking of reported values against climatology</u>: For the SHIP-report errors in the positions of the ships are very common. In this case it is of value to check some elements, e g the sea surface temperature, against climatological values.

7.2 Space continuity check

When analyzing manually the meteorologist rejects some observations and the decisions are generally based on a comparison with observations in the vicinity of the rejected observations. It is possible to use a similar technique when the analysis is performed numerically. I will briefly describe the space continuity check which is used in the analysis system of the Swedish Weather Service. When a certain observed value is to be checked, statistical interpolation from the observed values in the vicinity is used to determine the most likely value, to be compared with the observed value.

Let f_{it}^{OBS} be the observed value at station i at the time t, let f_{it}^{I} be the corresponding interpolated value and let E be the estimated mean square error of interpolation. If we assume random observational errors we will have:

$$\overline{(f_{it}^{OBS} - f_{it}^{I})^2} = \sigma_i^2 + E$$

where σ_i^2 denotes the variance of the (natural) observational error.

The following algorithm for checking the deviation between the observed and the interpolated values is used:

If $\left|f_{it}^{OBS} - f_{it}^{I}\right|$
$< K \cdot \sqrt{\sigma^2 + E}$ accept the observed value!
$> K \cdot \sqrt{\sigma^2 + E}$ Examine whether any of the observed values used for the interpolation is suspect (see below). If not - reject the observed value!

The value of the constant K has to be selected according to the acceptable risk of eliminating a correct observation. **If for simplicity we** assume that the difference between the interpolated and the true value (including the natural error) is normally distributed this risk can be computed:

If K = 1 the risk of rejecting a correct value is ~32%
" K = 2 "   "   "   "   "   "   "   "   ~ 4%
" K = 3 "   "   "   "   "   "   "   "   ~ 0.03%

The selection of a value for K will thus be based on a compromise between the risk of accepting an erroneous value and the risk of eliminating a correct value.

This compromise is unavoidable for all methods of statistical control. The difficulty lies in the fact that the most valuable information also is the information that is most difficult to check. Another complication with checking methods of this kind is that the observations which are used for the interpolation may also be erroneous. A straightforward application of the method may cause correct observations to be rejected by erroneous observations which are used to obtain the interpolated value. One way to avoid this difficulty is to repeat the interpolation during a one by one exclusion of the observations used for the interpolation.

The experience of this checking method is generally quite good at the Swedish Weather Service. The only problems encountered have been in the checking of surface pressure observations over the oceans. For these observations, a comparison has shown that multivariate interpolation is a stronger tool for space continuity check.

7.3   Manual intervention in the numerical analysis process

The amount of manual intervention as well as the procedure for manual intervention differ very much between weather services. Further, the procedures for manual intervention are seldomly described in scientific articles about the analysis schemes.

First of all it is obvious that the amount of manual intervention should be defined by the difficulty of the analysis task. It is quite clear that manual intervention is needed at the weather services of the Southern Hemisphere. From satellite pictures it is possible to identify the position and intensity of low pressure systems, and this visual information may be quantified and introduced into the numerical analysis process in the form of vorticity, wind or "bogus" pressure observations. It has been shown that this type of manual intervention has a significant positive impact on the quality of numerical forecasts of the Australian Weather Service.

With regards to short-range weather forecasts in the European-Atlantic area I would like to suggest the following principle for manual intervention in the numerical analysis process:

"In some situations it may be impossible to decide with numerical/statistical methods whether an observation is correct or not. Some observations have to be classified as "suspect". If the information content (by the computer) is considered to be crucial for the analysis result, it is considered advantageous to perform a manual checking by a meteorologist of the most valuable observations, provided the meteorologists has access to some independent information in the form of e g satellite pictures. Any quality control information or bogus data which is manually inserted should be carefully checked by the computer. Another necessary condition for such a cooperation between the meteorologist and the computer is a fast technical device for communication (e. g. a graphical display)."

9. CONCLUDING REMARKS

During a period of 30 years, operational schemes for numerical analysis of meteorological fields have been developed from simple two-dimensional univariate interpolation schemes to complex three-dimensional multivariate interpolation methods which are important elements of schemes for four-dimensional data assimilation. Today most weather services utilize statistical interpolation methods which are able to use results from numerical forecast models, past **experience** of statistical nature as well as knowledge of dynamical relationships. In addition, the properties of the observational errors are taken into account. Still it is appropriate to conclude that our present statistical interpolation schemes are rather undeveloped tools for numerical analysis of meteorological fields since they are based only on the historical performance of the forecast models. The interpolation errors are minimized over a large number of cases and this **certainly does not mean that the** interpolation weights are the best for each individual case. Several approaches to refine our statistical interpolation schemes have been suggested but have to be further developed:

- Utilization of stochastic-dynamic models to **also predict the forecast** covariances.
- Utilization of observed data (e g pressure tendencies) to adjust the forecast error covariances to the actual synoptic situation.

Statistical interpolation methods dominate today. Other methods also have advantages which need to be further explored. A synthesis of, for **example, statistical interpolation, spectral analysis and variational normal-mode initialization does not seem impossible to develop.**

While working with this huge problem, some other smaller problem areas also need further consideration:

- Analysis of the humidity field.
- Optimized algorithms for quality control of observational data.

LIST OF REFERENCES

Alaka, M.A. and Elvander, R.C.: Optimum interpolation from observations of mixed quality. Monthly Weather Review, Vol. 100, No. 8, pp. 612-624.

Atkins, M., 1974: The objective analysis of relative humidity. Tellus, Vol. 26, NO. 6, pp. 663-671.

Bengtsson, L. and Gustafsson, N., 1972: Assimilation of non-synoptic observations. Tellus, Vol. 24, pp. 383-399.

Bengtsson, L., 1975: 4-dimensional assimilation of meteorological observations, GARP Publication Series No. 15.

Bengtsson, L., 1980: On the use of a time sequence of surface pressures in four-dimensional data assimilation. Tellus, Vol. 32, pp. 189-197.

Bergman, K.H., 1979: Multivariate analysis of temperatures and winds using optimum interpolation. Monthly Weather Review, Vol. 107, pp. 1423-1444.

Bergthorsson, P. and Döös, B.R., 1955: Numerical weather map analysis. Tellus, Vol. 7, pp. 329-340.

Bertoni, E. and Lund, T.A., 1963: Space correlations of the height of constant pressure surfaces. J. Appl. Meteor., Vol. 2, MO 4.

Buell, E., 1972: Correlation functions of the height of constant pressure surfaces. J. Appl. Meteor., Vol. 11, pp. 51-59.

Cats, G.J., 1980: Construction of correlation matrices for the ECMWF analysis scheme. ECMWF working paper.

Cressman, G.P., 1959: An operational objective analysis system. Monthly Weather Review, Vol. 87, pp. 367-374.

Dixon, R., 1976: An objective analysis system using orthogonal polynomials. GARP WGNE Report No. 11, pp. 73-85.

Eliassen, A., 1954: Provisional report on calculation of spatial covariance and autocorrelation of the pressure field. Inct. Weather and Climate Res., Acad. Sci. Oslo, Rept. No. 5.

Flattery, T., 1971: Spectrql models for global analysis and forecasting. Proc. Sixth AWS Technical Exchange Conference. Air Weather Service Techn. Rep. 242, pp. 42-54.

Gandin, L.S., 1963: Objective analysis of meteorological fields. Leningrad. Hydromet. Press.

Gandin, L.S. and Tarasyuk, V.V., 1971: A complex method of checking the upper-air information. Meteorologija i Gidrologija, 1971, No. 5, pp. 3-9.

Ghil, M. and Balgovind, R., 1980: A Langevin equation for large-scale atmospheric flow. GARP WGNE Report No. 20, pp. 31-33.

Gilchrist, B. and Cressman, G.P., 1954: An experiment in objective analysis. Tellus, Vol. 6, pp. 97-101.

Hayden, C.M., 1976: Satellite reference level experiments with VTRR and the NMC global spectral analyses. GARP WGNE Report No. 11, pp. 57-72.

Hollett, S.R., 1975: Three-dimensional spatial correlations of PE forecast errors. M.S. thesis, Dept of Meteorology, McGill University.

Kaestner, 1974: Ein Verfahren zur numerischen Analyse der relativen Feuchte. Arch. Met. Geph. Biokl., Ser. A, Vol. 23,

Van Maanen, J., 1980: Objective analysis of the humidity field by the optimum interpolation method. GARP WGNE Report No. 20, pp. 38-40.

Panofsky, H.A., 1949: Objective weather map analysis. J. Met., Vol. 6, No.6, pp. 386-392.

Richardson, L.F., 1922: Weather prediction by numerical process. Dover Publ. Ins., New York 1965.

Rutherford, I.D., 1976: An operational three-dimensional multivariate statistical objective analysis scheme. GARP WGNE Report No. 11, pp. 98-121.

Sasaki, Y., 1958: An objective analysis based on the variational method. J. of Met. Soc. Japan, Vol. 36, pp. 77-88.

Thiebaux, H.J., 1975: Experiments with correlation representations for objective analysis. Mon. Wea. Rev., Vol. 103, pp. 617-627.

Thiebaux, H.J., 1976: Anisotropic correlation functions for objective analysis. Mon. Wea. Rev., Vol. 104, pp. 994-1002.

THE NORMAL MODE APPROACH TO THE INITIALIZATION PROBLEM

Roger Daley

National Center for Atmospheric Research[*]
Boulder, Colorado 80307, U.S.A.

ABSTRACT

The baroclinic primitive equation models used for short and medium range weather forecasting admit undesirable high frequency gravity waves as well as the desirable slow-moving Rossby modes. The gravity waves are excited by initial imbalances between the observed mass and wind fields and by inconsistencies between model and atmosphere. The initial separation of the meteorological noise (gravity waves) from the signal has been the long-time goal of model initialization procedures.

Normal mode initialization techniques have been found to be extremely proficient at separating and removing meteorological noise from the initial state. The present review will discuss the development of the technique from first principles and will give examples of successful applications.

---

[*] The National Center for Atmospheric Research is sponsored by the National Science Foundation

1. INTRODUCTION

Primitive equation models, unlike quasi-geostrophic models, generally admit high frequency gravity wave solutions, as well as the slower moving Rossby wave solutions. Although there are gravity waves in the atmosphere, they are generally of much smaller amplitude than the gravity waves which appear in primitive equation models. Since it is primarily the low frequency part of the flow which is of interest, model gravity modes are at best a nuisance and at worst can seriously compromise the forecast procedure. First of all, gravity waves generally require short computational timesteps; secondly, they can interfere seriously with very short period forecasts (< 12 hours); and thirdly, they can impair the precipitation and vertical motion calculations. Consequently, it is advantageous if they can be suppressed from PE model integrations.

Gravity wave oscillations arise primarily from initial imbalances between the wind and mass fields. These imbalances exist partially because the observed or analyzed mass and wind fields contain error and partially because the model equations do not exactly describe the atmosphere. Gravity wave oscillations can be controlled to some extent by the addition of time or space dissipation terms to the model equations. However, the primary way of suppressing gravity waves is by balancing the initial state through an initialization procedure.

Classical static initialization procedures such as the linear and non-linear balance equations and variational techniques have been widely used. Dynamic initialization procedures, in which the model equations are integrated forward and backward in time with large damping factors, have succeeded in suppressing many of the gravity waves. However, like the classical static techniques they have not worked well for the larger scales or in the tropics.

The use of model normal modes for initialization is attractive because the normal modes each have an associated frequency and one can, in principle, suppress only the high frequency gravity modes. Flattery (1970) developed an analysis/initialization procedure based on the Laplace tidal equations. Williamson (1976) and Williamson and Dickinson (1976) found the normal modes of the NCAR general circulation model. Their initialization procedure simply set the initial amplitudes of the high frequency gravity modes equal to zero. This procedure was only partially successful because the initial projection of the non-linear and forcing terms of the equations onto these high frequency modes rapidly re-excited them. Machenhauer (1977) and Baer (1977) independently overcame this problem with non-linear normal mode initialization which took some account of the non-linearities. Their procedures have been applied to several different models with considerable success.

In the following sections, we will progress logically through the construction of the model normal modes, their properties and their use in initialization procedures. We shall then consider the relationship of normal mode initialization to quasi-geostrophic theories and outline the theory of the slow manifold. We will finish with some examples of the successful application of this technique. For a more comprehensive review of normal mode initialization, see Daley (1981).

2. THE FORECAST MODEL

The first step in normal mode initialization is to find the free normal modes of the forecast model. (We shall distinguish between the free and forced normal mode problems, later.) The model we shall use here is a baroclinic primitive equations model in pressure coordinates. This is a slightly simpler model than the usual sigma coordinate model, but the derivation of the normal modes is virtually the same. The normal mode problem for a pressure coordinate model was first solved by Flattery (1967), but we shall use an approach similar to that of Kasahara and Puri (1980). The equations of motion, thermodynamic and continuity equations for this model are

$$\frac{\partial \underline{v}}{\partial t} + \underline{k} \times f\underline{v} + \nabla \Phi = R_{\underline{v}} \tag{1}$$

$$\frac{\partial}{\partial t} \frac{\partial \Phi}{\partial P} + \frac{R \gamma^* \omega}{P} = R_\Phi \tag{2}$$

$$\frac{\partial \omega}{\partial P} + \nabla \cdot \underline{v} = 0 \tag{3}$$

where

$$R_{\underline{v}} = -\underline{k} \times \zeta\underline{v} - \tfrac{1}{2}\nabla \underline{v}\cdot\underline{v} - \omega\frac{\partial \underline{v}}{\partial P}, \qquad (4)$$

$$R_\phi = -\underline{v}\cdot\nabla\frac{\partial \phi}{\partial P} - \frac{R\gamma'\omega}{P}.$$

$$\zeta = \underline{k}\cdot\nabla \times \underline{v},$$

$$\gamma = -\frac{T}{\theta}\frac{\partial \theta}{\partial P} = \frac{RT}{C_p P} - \frac{\partial T}{\partial P},$$

γ^* = horizontal average of γ, γ' = deviation,

R, C_p, P, ϕ are the gas constant, specific heat at constant pressure, pressure and geopotential,

$\omega = \frac{dP}{dt}$ and $\underline{k}$ is the vertical unit vector.

The vertical boundary conditions for this model are $\omega = 0$ at $P = 0$, and $w = \frac{dz}{dt} = 0$ at the ground ($z = 0$).

3. LINEARIZATION

The next step in generating the normal modes is to define a basic state and then linearize the model about it. We will define a very simple basic state - no flow and with a basic state static stability γ^* which is a function only of P. The structures and frequencies of the high frequency gravity modes are largely unaffected by the imposition of a more realistic basic state (see Kasahara, 1980). Since, in the initialization problem, it is the high frequency modes which are of primary concern, a basic state at rest is sufficient.

We shall write the model equations in spherical polar coordinates with the linearized terms on the left hand side and the non-linear terms on the right hand side:

$$\frac{\partial u}{\partial t} - 2\Omega \sin\phi\, v + \frac{1}{a\cos\phi}\frac{\partial \phi}{\partial \lambda} = R_u \qquad (5)$$

$$\frac{\partial v}{\partial t} + 2\Omega \sin\phi\, u + \frac{1}{a}\frac{\partial \phi}{\partial \phi} = R_v \qquad (6)$$

$$\frac{\partial}{\partial t}\frac{\partial \phi}{\partial P} + \frac{R\gamma^*\omega}{P} = R_\phi \tag{7}$$

$$\frac{\partial \omega}{\partial P} + \frac{1}{a\cos\phi}\left[\frac{\partial u}{\partial \lambda} + \frac{\partial}{\partial \phi}v\cos\phi\right] = 0 \tag{8}$$

where a = earth radius, Ω = earth rotation rate, λ, ϕ = longitude and latitude, and u, v = zonal and meridional velocity components. The linearized forms of the equations are obtained by setting $R_u = R_v = R_\phi = 0$.

We can eliminate the variable ω by differentiating the linearized form of equation (7) with respect to P and introducing equation (8).

$$\frac{\partial}{\partial t}\frac{\partial}{\partial P}\frac{P}{R\gamma^*}\frac{\partial \phi}{\partial P} - \frac{1}{a\cos\phi}\left[\frac{\partial u}{\partial \lambda} + \frac{\partial}{\partial \phi}v\cos\phi\right] = 0. \tag{9}$$

4. THE VERTICAL STRUCTURE EQUATION

We will now attempt to separate the horizontal and vertical dependence of the linearized equations (5, 6 and 9) by assuming that the dependent model variables u, v and ϕ can be written as follows:

$$\begin{bmatrix} u \\ v \\ \phi \end{bmatrix} = \begin{bmatrix} \hat{u}(\lambda, \phi, t) \\ \hat{v}(\lambda, \phi, t) \\ \hat{\phi}(\lambda, \phi, t) \end{bmatrix} Z(P) \tag{10}$$

where $Z(P)$ gives the vertical structure and $\hat{u}$, $\hat{v}$ and $\hat{\phi}$ give the horizontal and temporal structure. Using the normal techniques of separation of variables, Eqs. 5, 6 and 9 (with $R_u = R_v = R_\phi = 0$) can be written as three horizontal equations

$$\frac{\partial \hat{u}}{\partial t} - 2\Omega\sin\phi\,\hat{v} + \frac{1}{a\cos\phi}\frac{\partial \hat{\phi}}{\partial \lambda} = 0 \tag{11}$$

$$\frac{\partial \hat{v}}{\partial t} + 2\Omega\sin\phi\,\hat{u} + \frac{1}{a}\frac{\partial \hat{\phi}}{\partial \phi} = 0 \tag{12}$$

$$\frac{\partial \hat{\phi}}{\partial t} + \frac{g\bar{H}}{a\cos\phi}\left[\frac{\partial \hat{u}}{\partial \lambda} + \frac{\partial}{\partial \phi}\hat{v}\cos\phi\right] = 0 \tag{13}$$

and a vertical structure equation which is a function of P only

$$\frac{\partial}{\partial P} \frac{P}{R\gamma^*} \frac{\partial Z}{\partial P} + \frac{Z}{g\tilde{H}} = 0 \qquad (14)$$

where g is the gravitational constant, $\tilde{H}$ is called the equivalent depth and $(g\tilde{H})^{-1}$ is the separation constant. We note that Eqs. (11 - 13) are analagous to the linearized shallow water equations of a fluid with mean depth $\tilde{H}$, a fact first noted by Taylor (1936).

5. THE HORIZONTAL STRUCTURE EQUATION

The horizontal equations can be further separated by assuming an exponential behavior in time and longitude. Thus

$$\begin{bmatrix} \hat{u} \\ \hat{v} \\ \hat{\phi} \end{bmatrix} = \begin{bmatrix} \hat{u}^m \\ i\ \hat{v}^m \\ 2\Omega\ \hat{\phi}^m \end{bmatrix} \exp(im\lambda - 2\Omega i\sigma t) \qquad (15)$$

where m is the zonal wavenumber, σ is a non-dimensional frequency and $i = \sqrt{-1}$. Substituting expansion (15) into Eqs. (11 - 13) gives

$$\sigma \hat{u}^m = -\sin\phi\ \hat{v}^m + \frac{m\ \hat{\phi}^m}{a\cos\phi} \qquad (16)$$

$$\sigma \hat{v}^m = -\sin\phi\ \hat{u}^m - \frac{1}{a} \frac{\partial \hat{\phi}^m}{\partial \phi} \qquad (17)$$

$$\sigma \hat{\phi}^m = \frac{g\tilde{H}}{4\Omega^2 a\cos\phi} \left[m\hat{u}^m + \frac{\partial}{\partial \phi} \hat{v}^m \cos\phi \right] \qquad (18)$$

The elimination of $\hat{u}^m$ and $\hat{v}^m$ between Eqs. (16, 17 and 18) leads to the so-called horizontal structure equation

$$H(\hat{\phi}^m) + \frac{4\Omega^2 a^2}{g\tilde{H}}\ \hat{\phi}^m = 0 \qquad (19)$$

where

$$H = \left\{ \frac{1}{\cos\phi} \frac{\partial}{\partial \phi} \left[\frac{\cos\phi}{(\sigma^2 - \sin^2\phi)} \frac{\partial}{\partial \phi} \right] + \frac{1}{(\sigma^2 - \sin^2\phi)} \left[\frac{m}{\sigma} \frac{(\sigma^2 + \sin^2\phi)}{(\sigma^2 - \sin^2\phi)} - \frac{m^2}{\cos^2\phi} \right] \right\}$$

is known as the horizontal structure operator.

Equation (14) - the vertical structure equation and Equation (19) are eigenvalue problems. The eigenvalue for the vertical structure equation is $(g\tilde{H})^{-1}$ while the eigenvalue for the horizontal problem is the frequency σ. We shall distinguish here between the forced and free normal mode problems. In the forced problem, a forcing function and forcing frequency σ are imposed. In this case the horizontal equation (with σ specified) is solved to find the equivalent depth $\tilde{H}$ and then this $\tilde{H}$ is inserted in equation (14) to find the vertical structure.

It is the free normal mode problem which is of interest here. In the free case, the vertical structure equation is solved first to obtain a set of vertical eigenvectors each with associated equivalent depth $\tilde{H}$. We then have one horizontal eigenvalue problem to solve for each equivalent depth. In principle, with $\tilde{H}$ thus specified, we can solve the horizontal structure equation (14) obtaining a set of horizontal eigenvectors, each with an associated eigenfrequency σ. In practice, it is easier to solve the eigenvalue problem defined by equations (16 - 18) - which is equivalent to equation (19). This is because the eigenvalue σ appears in a complicated form in euqation (19).

6. BOUNDARY CONDITIONS

Before going on to discuss the solution of the vertical and horizontal structure equations, we must first discuss the spatial boundary conditions for these equations.

In the horizontal, we have assumed spherical polar geometry and there are no boundary conditions. In the case of a limited domain model, we would of course need boundary conditions for equations (16 - 18).

In the vertical, we need boundary conditions on Z at the top and bottom of the atmosphere. We will make use of the vertical boundary conditions discussed earlier, $\omega = 0$ at $P = 0$, and $w = 0$ at the ground.

At the top, we have from the thermodynamic equation (7) that

$$\frac{\partial Z}{\partial P} = 0 \quad \text{at } P = 0 \tag{20}$$

We can find the appropriate condition on Z at the ground as follows. We know that at the earth's surface (in the absence of topography)

$$w = \frac{dz}{dt} = 0$$

Linearizing and using the hydrostatic equation, we find

$$\frac{\partial \Phi}{\partial t} - \frac{\omega R T^*}{P_0} = 0$$

where P_0 is the mean pressure at the ground, from which it can be shown that

$$\frac{\partial Z}{\partial P} + \frac{\gamma^* Z}{T^*} = 0 \tag{21}$$

where T^* = horizontally averaged temperature at the ground.

7. VERTICAL STRUCTURE FUNCTIONS

The vertical structure equation (14) and horizontal structure equations (16 - 18) are converted into algebraic eigenvalue problems by discretization. We shall concentrate on the discretization of the vertical structure equation, as it is more straightforward.

Most NWP models use a gridpoint representation in the vertical. Equation (14) with boundary conditions (21 and 22) would be solved using exactly the same vertical levels and finite-differencing used in the models. Imagine that the model uses 2nd order centered differencing and has L vertical levels with $\ell = 1$, L ranging from the top to the bottom of the models. We would then find that equation (14) would have a finite difference representation for an arbitrary level ℓ as follows:

$$\frac{2}{(\Delta P_\ell + \Delta P_{\ell+1})} \left[B_{\ell+\frac{1}{2}} \frac{Z(\ell+1) - Z(\ell)}{\Delta P_{\ell+1}} - B_{\ell-\frac{1}{2}} \frac{Z(\ell) - Z(\ell-1)}{\Delta P_{\ell-1}} \right] + \frac{Z(\ell)}{g\tilde{H}} = 0$$

where $\tag{22}$

ΔP_ℓ = difference of pressure between levels,

$Z(\ell)$ = value of Z at level ℓ,

$B_{\ell+\frac{1}{2}} = P_{\ell+\frac{1}{2}}/R\gamma^*_{\ell+\frac{1}{2}}$ and ½ indicates half level.

All other levels would be similar except $\ell = 1$ and $\ell = L$ where the boundary conditions (20 - 21) would be used. This particular discretization is that used by Kasahara and Puri (1980). Other finite difference or finite element discretizations are given in Kasahara (1976), Daley (1979), and Temperton and Williamson (1979).

After discretization, equation (14) with boundary conditions (20 - 21) appears as the following algebraic eigenvalue problem:

$$\underline{\underline{A}} \, \underline{Z} + (g\tilde{H})^{-1} \, \underline{Z} = 0 \qquad (23)$$

where

$\underline{Z}$ is the column vector of $Z(\ell)$,

$\underline{\underline{A}}$ is the matrix of finite difference coefficients defined in equation (22).

Equation (23) can be solved by standard algebraic eigenvalue techniques to give L eigenvectors $Z_k(p)$, $k = 1$ to L, each with an associated eigenvalue (equivalent depth) $\tilde{H}_k$.

The vertical eigenvectors and associated equivalent depths for a typical atmospheric model (Kasahara and Puri, 1980) can be seen in Figure 1. There are 9 vertical levels and thus 9 eigenmodes. The vertical levels are indicated roughly as a function of P/P_o. The modes are ordered by decreasing equivalent depth. The gravest vertical mode (equivalent depth = 9750 m) is called the external mode, the others are called first internal, second internal, etc. The external mode has very little vertical structure. With decreasing equivalent depth the modes have more and more of their structure near the ground.

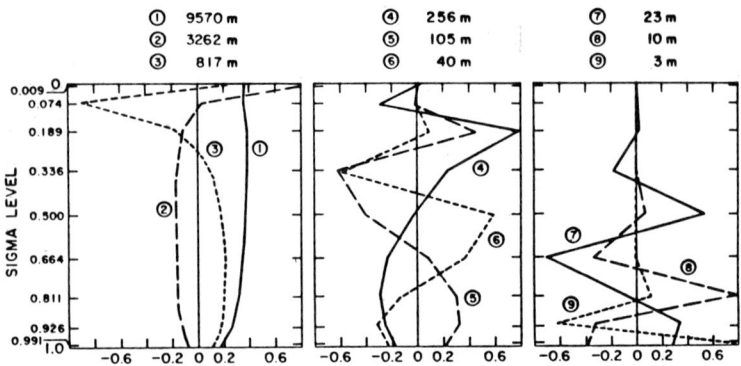

Figure 1 - after Kasahara and Puri (1980)

8. HORIZONTAL STRUCTURE FUNCTIONS

As mentioned in Section 5, the next step in the free normal mode problem is to take each of the L equivalent depths $\tilde{H}_k$, k = 1 to L, and solve the corresponding horizontal eigenproblem (19) or equations (16 - 18) for each $\tilde{H}_k$.

The horizontal eigenproblem can be turned into an algebraic eigenvalue problem by discretization. The traditional method for the spherical case is to expand the dependent variables in a spherical harmonic series. This leads to an algebraic eigenvalue problem for each zonal wavenumber. The solutions to this problem in the limit of an infinite spherical harmonic expansion are known as the horizontal structure functions or Hough functions (Hough, 1898; Longuet-Higgins, 1968). With each horizontal structure functions is an associated real eigenfrequency.

In the present case, however, we are attempting to find the free normal modes of atmospheric models which have only a finite number of degrees of freedom. Therefore, we are not interested in the actual Hough functions themselves, but discrete approximations to them.

In the case of a spectral model in which a truncated spherical harmonic expansion is used for the horizontal discretization, the horizontal structure functions would differ only slightly from the true Hough functions on the large-scale, but would differ appreciably on the small scale. In the case of a finite difference model, a finite difference approximation consistent with the discretization of the model would be used in the discretization of equations (16 - 18). The resulting horizontal structure functions would again be similar to true Hough functions in the large scale, but differ for the small scale. For limited area models with lateral boundaries, there would also be boundary modes.

Let us suppose that our model has M zonal (east-west) degrees of freedom and N meridional (north-south) degrees of freedom. Thus, there are for each vertical mode k, MN horizontal degrees of freedom for each of the 3 dependent variables (u, v, Φ). Thus, there will be 3 MN horizontal structure functions for each vertical mode, each having a u, v, Φ component denoted below

$$\hat{u}_n^k(\lambda, \phi), \quad \hat{v}_n^k(\lambda, \phi), \quad \hat{\Phi}_n^k(\lambda, \phi) \quad 1 \leq n \leq MN$$

where n indicates the horizontal mode number. With each of the 3 MN horizontal structure functions is an associated eigenfrequency σ_n^k.

Now for the spherical case the modes can be classified by symmetry. In the symmetric case the modes $\hat{u}_n^k$, $\hat{\Phi}_n^k$ are symmetric with respect to the equator, while $\hat{v}_n^k$ is anti-symmetric. The symmetric modes would be the appropriate set for a hemispheric model. There is also an anti-symmetric set in which $\hat{u}_n^k$, $\hat{\Phi}_n^k$ are anti-symmetric while $\hat{v}_n^k$ is symmetric.

The horizontal structure functions can also be classified by their eigenfrequencies. There are basically two classes of modes. The first class consists of high frequency eastward and westward propagating gravity modes while the second class consists of low frequency westward propagating Rossby waves.

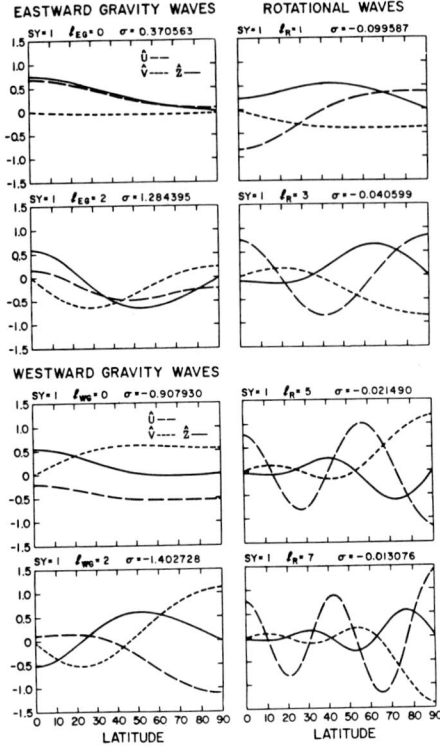

Figure 2 - after Kasahara (1976)

In Figure 2 is shown an example of meridional structure functions for zonal wavenumber m = 1 for the eastward and westward propagating gravity modes and the Rossby modes. The modes shown are all symmetric and the non-dimensional eigenfrequency σ is also indicated. The $\hat{u}$, $\hat{v}$, and $\hat{\phi}$ structures are all shown. This figure from Kasahara (1976) shows the horizontal structure of true Hough modes, but one could imagine that the structure of these large-scale modes would not be much affected by discretization. The equivalent depth in this example is 10 km, corresponding to an external mode.

It can be noticed from Figure 2 that the magnitudes of the eigenfrequencies for the gravity modes are much larger than for the Rossby modes. For smaller equivalent depths, however, the absolute values of the eigenfrequencies for all modes tend to decrease. This effect can be seen in Table 1. In this table are plotted the non-dimensional eigenfrequencies for zonal wavenumber 1 for 2 equivalent depths, 10 km and .1 km. The frequencies of the 5 gravest symmetric gravity and Rossby modes are shown. The frequencies do not quite agree with those shown in Figure 2 because they were obtained for a slightly different model.

Table 1

Non-dimensional Eigenfrequencies

Zonal Wavenumber 1

	10 km			.1 km	
Eastward Gravity	Westward Gravity	Rossby	Eastward Gravity	Westward Gravity	Rossby
.3693	-.9066	-.0994	.0339	-.3121	-.0112
1.2810	-1.3996	-.0406	.3246	-.4770	-.0049
1.9514	-1.9902	-.0215	.4833	-.5931	-.0032
2.6074	-2.6267	-.0131	.5980	-.6859	-.0024
3.2676	-3.2792	-.0087	.6903	-.7636	-.0019

It should be noted that some of the gravity modes for the .1 km equivalent depth have frequencies almost as small as the Rossby modes with equivalent depth 10 km.

For the spherical model, the Rossby modes for zonal wavenumber 0 are degenerate in that they have zero eigenvalues. Kasahara (1978) has considered this problem.

9. NORMAL MODE PROPERTIES

We will designate a particular normal mode using the following notation:

$$\hat{\Pi}_n^k(\lambda, \phi, p) = \begin{bmatrix} \hat{u}_n^k(\lambda, \phi) \\ \hat{v}_n^k(\lambda, \phi) \\ \hat{\phi}_n^k(\lambda, \phi) \end{bmatrix} Z_k(P) \tag{24}$$

for each $\tilde{H}_k$ and eigenfrequency σ_n^k.

The $\hat{\Pi}_n^k$, being solutions to a linear problem, have an arbitrary amplitude. They are orthogonal, however, and can be normalized. We will write the orthonormality condition using inner product notation.

$$<\hat{\Pi}_n^k \cdot \hat{\Pi}_\ell^j> = \delta_k^j \, \delta_n^\ell \tag{25}$$

where δ_j^k is the Kronecker delta. Equation (25) follows from the fact that the horizontal and vertical structure functions are individually orthogonal. Thus

$$\int_P Z_k(P) \, Z_j(P) \, dP = \delta_k^j \qquad (26)$$

$$\int_\lambda \int_\phi \left[\frac{g\tilde{H}_k}{4\Omega^2} \left(\hat{u}_n^k \, \hat{u}_\ell^k + \hat{v}_n^k \, \hat{v}_\ell^k \right) + \hat{\Phi}_n^k \, \hat{\Phi}_\ell^k \right] \cos\phi \, d\phi \, d\lambda = \delta_n^\ell$$

We have written these orthogonality conditions as if the vertical and horizontal structure functions were analytic. In general, these structure functions will only be defined at gridpoints and equations (25 and 26) would be replaced by numerical quadrature expressions consistent with the model discretization.

Let us define Π to be a vector of arbitrary wind and geopotential fields

$$\Pi = \begin{bmatrix} u(\lambda, \phi, P) \\ v(\lambda, \phi, P) \\ \Phi(\lambda, \phi, P) \end{bmatrix} \qquad (27)$$

Then Π can be expanded in a series of normal mode functions in the same way one expands a field in a Fourier series. Thus

$$\Pi = \sum_n \sum_k x_n^k \, \hat{\Pi}_n^k(\lambda, \phi, P) \qquad (28)$$

where

$$x_n^k = \langle \Pi \cdot \hat{\Pi}_n^k \rangle \qquad (29)$$

x_n^k is called the normal mode expansion coefficient. It can be obtained in equation (29) by projecting the data (u, v, Φ) onto the horizontal and vertical structure functions appropriate to that mode and making use of the orthonormality conditions (26).

10. NORMAL MODE FORM OF MODEL EQUATIONS

We can use equation (29) to write the linearized form of the model equations (5 - 8) in normal mode form. Thus, with $R_u = R_v = R_\phi = 0$, we have

$$\dot{x}_n^k + 2\Omega i \, \sigma_n^k \, x_n^k = 0, \quad \text{all } k, n \tag{30}$$

where x_n^k is the expansion coefficient corresponding to the normal mode defined by (kn, n). This equation is produced from the linearized form of equations (5 - 8) simply by multiplying by the appropriate horizontal and vertical structure functions and integrating over the atmosphere as in equation (28).

Let us define X to be the vector of all normal mode expansion coefficients, i.e. each x_n^k is a member of X. Our real goal, here, is to find the normal mode form of the full non-linear equations (R_u, R_v, $R_\phi \neq 0$). Let us define R_Π to be the column vector (analogous to Π in equation 27) of R_u, R_v, R_ϕ. Thus,

$$R_\Pi = \begin{bmatrix} R_u(\lambda, \phi, P) \\ R_v(\lambda, \phi, P) \\ R_\phi(\lambda, \phi, P) \end{bmatrix} \tag{31}$$

R_Π can be projected onto the normal modes through an expression analogous to equation (29). Thus the full non-linear form of equations (5 - 8) can be written in normal mode form as

$$\dot{x}_n^k = -2\Omega i \, \sigma_n^k \, x_n^k + R_n^k(X) \tag{32}$$

where

$$R_n^k(X) = \langle R_\Pi \cdot \hat{\Pi}_n^k \rangle \tag{33}$$

Note that since R_u, R_v, R_ϕ are non-linear terms, R_n^k is therefore a function of all the normal mode expansion coefficients and not simply of x_n^k. Thus we write R_n^k as a function of the column vector X defined earlier. We can also write formally an equation for the vector of normal mode expansion coefficients.

$$\dot{X} = -2\Omega i \, \Lambda_x X + R_x(X) \tag{34}$$

where Λ_x is a diagonal matrix whose elements are the non-dimensional eigenfrequencies σ_n^k and $R_x(X)$ is a short-hand notation for the projection of the non-linear terms on each of the normal modes in turn.

Equation (34) is a symbolic equation and would have the same form, no matter what the original model equations had been.

11. FAST AND SLOW EQUATIONS

The next step is to divide the set of normal modes up into fast modes and slow modes on the basis of their frequency. For many applications, it would be sufficient to define the fast modes simply to be the set of all gravity modes and the slow modes to be the set of all Rossby modes. Other applications require a more subtle distinction between fast and slow modes.

We will refer to the set of fast modes as Z and the set of slow modes as Y. Thus X = Z + Y. We can then re-write the normal mode form of the equation (34) in terms of a fast equation and a slow equation.

$$\dot{Z} = -2\Omega i \, \Lambda_Z Z + R_Z(Z, Y) \tag{35}$$

$$\dot{Y} = -2\Omega i \, \Lambda_Y Y + R_Y(Z, Y) \tag{36}$$

The non-linear projection onto the fast modes R_Z is a function of both fast and slow modes, and similarly for R_Y. In general, the frequencies Λ_Y are small compared with the frequencies Λ_Z.

12. LINEAR INITIALIZATION

We are now ready to consider the initialization problem. Suppose the model equations were the linearized form of equations (5 - 8), i.e. with $R_u = R_v = R_\phi = 0$. Suppose we wish to eliminate high frequency oscillations (which we identify with the fast modes Z) from the model integration. The fast equation for this model, then, is

$$\dot{Z} = -2\Omega i \, \Lambda_Z Z.$$

To eliminate fast oscillations for all time, it suffices at initial time to set

$$Z = 0. \tag{37}$$

However, if we have the full non-linear equations (R_u, R_v, $R_\phi \neq 0$) then the appropriate fast equation is equation (35). In this case, setting $Z = 0$ at $t = 0$ will not suppress fast oscillations for all time as they will be re-excited by the non-linear term R_Z. This linear procedure was applied to the non-linear shallow water equations by Williamson (1976) and it was not capable of eliminating the fast oscillations.

13. NON-LINEAR INITIALIZATION - MACHENHAUER'S SCHEME

In an experiment performed with the non-linear shallow water equations, Machenhauer (1977) plotted separately the linear and non-linear contributions to the time tendencies of particular fast modes during the integration of the model. Thus, for a particular fast mode z from the set Z he would plot $2\Omega i \sigma_z z$ and $R_z(Z, Y)$ as a function of time. We show in Figure 3 a slightly modified version of Machenhauer's (1977) result.

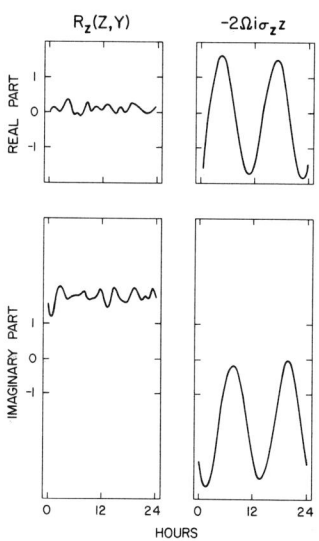

Figure 3 - after Machenhauer (1977)

In this diagram is shown the time behavior of a particular fast mode (the gravest gravity mode of zonal wavenumber 0). Both real and imaginary parts are shown. The curves on the left indicate the non-linear term while those on the right indicate the linear term. It can be seen that the linear oscillations are large, and in fact, have the same frequency as would be given by the linearized equations. The non-linear term, on the other hand, has only a very low amplitude high frequency oscillation.

Furthermore, in a time-averaged sense, the linear and non-linear terms nearly sum to zero.

This suggested to Machenhauer (1977) a scheme whereby the linear and non-linear terms would be balanced at initial time. This would be done by setting the initial time tendencies $\dot{Z}$ in equation (35) equal to zero at time t = 0. This would give rise to a diagnostic equation which could be solved for Z. Note, however, that it is a non-linear equation in Z and requires iteration to convergence. Machenhauer's scheme does not change the slow modes Y. The scheme is as follows.

Step 1 is to apply linear initialization. The subscript indicates the iteration step.

$$Z_0 = 0, \quad Y_0 = Y \tag{38}$$

Step 2 is to set $\dot{Z} = 0$ in equation (35)

$$Z_1 = \frac{R_Z(Z_0, Y_0)}{2\Omega i \, \Lambda_Z} \tag{39}$$

The next step is to repeat Step 2 except using Z_1 instead of Z_0. Thus

$$Z_2 = \frac{R_Z(Z_1, Y_0)}{2\Omega i \, \Lambda_Z} \tag{40}$$

This procedure would be repeated until convergence when the Z on the left hand side equalled the Z on the right hand side. The final values of the fast modes Z we will denote Z_B, where the subscript B stands for balanced. Thus

$$Z_B = \frac{R_Z(Z_B, Y_0)}{2\Omega i \, \Lambda_Z} \tag{41}$$

Now a non-linear iteration procedure such as this can only be expected to converge when the non-linear terms R_Z are small compared to the linear terms. This will be the case when the frequencies Λ_Z are large compared to the slow mode frequencies Λ_Y.

14. NON-LINEAR INITIALIZATION - THE BAER-TRIBBIA SCHEME

Completely independently of Machenhauer, Baer and Tribbia developed a non-linear normal mode initialization scheme. This work is covered in Baer (1977) and Baer and Tribbia (1978). See also Ballish (1979).

The initialization scheme of Baer and Tribbia is more complex and difficult to use in practice than that of Machenhauer. However, it is also more general and powerful than Machenhauer's scheme.

The normal mode form of the model equation (39) has been derived from the dimensional form of the model equations (1 - 3). The Baer-Tribbia methodology can be best understood by considering the non-dimensional form of the original model equations. Thus, if we had first non-dimensionalized equations (1 - 3) using appropriate length and time scales, equation (34) could be rewritten in the following form:

$$\frac{\partial X}{\partial t} = -2\Omega i \, \Lambda_X \, X + \varepsilon \, R_X(X) \tag{42}$$

where ε is a parameter which describes the strength of the non-linearity in the system. If the original model equations had been the shallow water equations, for example, then ε would correspond to a Rossby number (ratio of the non-linear to the Coriolis terms). For the planetary scales ε is generally taken to be small $O(.1)$. The analogues of the fast and slow equations (35 and 36) using this non-dimensionalization are

$$\frac{\partial Z}{\partial t} = -2\Omega i \, \Lambda_Z \, Z + \varepsilon \, R_Z(Z, Y) \tag{43}$$

$$\frac{\partial Y}{\partial t} = -2\Omega i \, \varepsilon \, \Lambda_Y \, Y + \varepsilon \, R_Y(Z, Y) \tag{44}$$

In these equations, all variables are $O(1)$ except ε. Note that an ε appears in the linear part of equation (44) because the frequencies Λ_y and Λ_z are assumed here to be of order unity whereas in equations (35 and 36) they were not.

We assume Z and Y are expanded in a power series in ε. Thus

$$Z = Z_0 + Z_1 \epsilon + Z_2 \epsilon^2 \ldots$$
$$Y = Y_0 + Y_1 \epsilon + Y_2 \epsilon^2 \ldots \quad (45)$$

We also assume that local time changes $\frac{\partial}{\partial t}$ can be represented by the sum of local time changes due to a fast time scale τ and a slow time T, namely,

$$\frac{\partial}{\partial t} = \frac{\partial}{\partial \tau} + \epsilon \frac{\partial}{\partial T} \quad (46)$$

We can now rephrase the initialization problem as follows. Without changing Y, how can we adjust Z at initial time, so that there are no fast time oscillations for all time. This requires that $\frac{\partial}{\partial \tau} = 0$.

We will omit the details of the analysis which can be seen in Baer (1977) and Baer and Tribbia (1978) and give the results directly. Thus to second order in ϵ, the Baer-Tribbia scheme gives

$$Z_0 = 0, \quad Y_0 = Y \quad (47)$$

$$Z_1 = \frac{R_Z(Z_0, Y_0)}{2\Omega i \, \Lambda_Z} \quad (48)$$

$$Z_2 = \frac{1}{2\Omega i \, \Lambda_Z} \left[R_Z(Z_1, Y_0) - \frac{\partial Z_1}{\partial T} \right] \quad (49)$$

where a small approximation has been made in the second term of equation (49).

The first pass is simply linear initialization. The second pass is equivalent to the second pass through Machenhauer's scheme. The third pass is similar to the third pass through Machenhauer's scheme except for the addition of an extra term. Machenhauer's scheme is therefore not consistent to second order in ϵ, as the neglected term $\frac{\partial Z_1}{\partial T}$ may be of the same order as the retained terms.

The practical difficulty with the Baer-Tribbia technique is in the evaluation of the term $\frac{\partial Z_1}{\partial T}$ in equation (49). It is possible, however, to evaluate this term by differentiating equation (48) with respect to T and evaluating the result numerically (see Tribbia, 1979).

15. THE SLOW MANIFOLD

Leith (1980) developed a graphical display which aids in the understanding of the initialization problem. The model normal modes can be represented in a multi-dimensional vector space. The Rossby and gravity modes each represent a subset or manifold of this vector space. In addition, Leith (1980) introduces the slow manifold which is defined to be the locus of all model states which are evolving slowly in time (i.e. have no fast time oscillations).

We can consider an approximation to the slow manifold by considering the locus of all points where $\dot{Z} = 0$ (equation 41). Figure 4, which is a modified version of Figure 1 of Leith (1980) is a simple two-dimensional illustration of the slow manifold concept. The amplitude of the Rossby modes Y is the abscissa while the gravity mode amplitude Z is the ordinate. We will refer to the abscissa as the Rossby manifold and to the ordinate as the gravity manifold. Any model state on the Rossby manifold could be said (loosely) to be in geostrophic balance. The slow manifold M is the locus of all model states where $Z = Z_B$. <u>The curved form of the slow manifold is a reflection of the non-linear nature of the balance. Any point on the slow manifold represents a model state which is evolving slowly in time</u>.

As $(Z, Y) \to 0$ the non-linear term $(R_z, R_y) \to 0$ and the slow manifold becomes coincident with the Rossby manifold. Conversely, as Y becomes large, the slow and Rossby manifolds diverge. Thus, a high amplitude, but slowly evolving model state would be far from geostrophic.

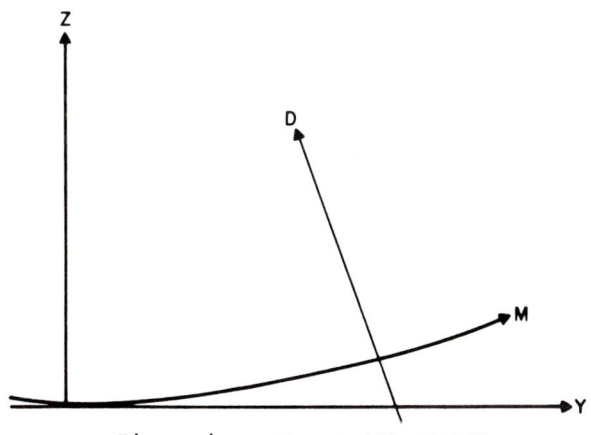

Figure 4 - after Leith (1980)

A particular point in this multi-dimensional space can have only a single spatial configuration for each of the dependent variables (geopotential, wind, etc.) of the model. Suppose, however, that the spatial structure of one of the model variables (geopotential, say) were held fixed while the other variables were allowed to vary. The locus of model states which satisfy these conditions, Leith (1980) refers to as a data manifold. This manifold is indicated in Figure 4 by the line D.

16. CONSTRAINED INITIALIZATION

The initialization procedures we have described up to this point (both Machenhauer and Baer-Tribbia formulations) can be characterized as <u>unconstrained</u> initialization procedures. We will now consider another class of initialization techniques, which will be referred to as <u>constrained</u> initialization procedures.

The concept of the slow manifold is very useful in graphically illustrating these different types of initialization procedures. Figure 5 is a schematic slow manifold diagram which will be used to illustrate constrained and unconstrained initialization. The Rossby manifold Y, slow manifold M, gravity manifold Z, and data manifold D are as in Figure 4.

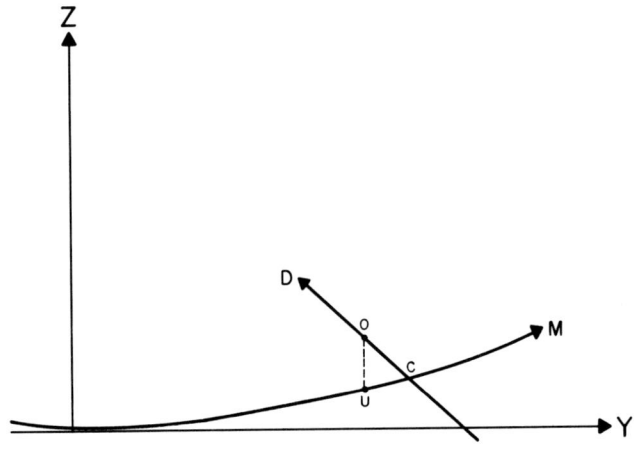

Figure 5

Suppose we have a model state at point O which lies on a data manifold D. We note that this point does not lie on the slow manifold, so that if used as an initial state for the model, gravity waves of magnitude proportional to the distance of O from M would be excited.

Now if we apply the unconstrained Machenhauer or Baer-Tribbia schemes discussed earlier, we can find an initial state on the slow manifold which will not excite gravity waves. In these schemes, it is assumed that the Rossby mode projection Y remains fixed. This corresponds to finding the point U (unconstrained) which is the intersection of the slow manifold and a vertical straight line passing through O.

To fix ideas, let us suppose that the data manifold D is a geopotential manifold (i.e. the spatial structure of the geopotential field would be invariant along the line D). Furthermore, we suppose that our observations of the geopotential are everywhere very accurate, but that our wind observations are very poor. If we performed unconstrained initialization, thus arriving at the point U, the original geopotential data would no longer be fitted. Since we had great faith in our original geopotential observations, our initialized state (U) would clearly be unsatisfactory.

A much more reasonable initial state could be obtained by finding the intersection of the data manifold D and the slow manifold M. In the present case, this would imply that the original geopotential observations were fitted and yet no high frequencies would be excited in subsequent model integrations. This state is indicated by the point C (constrained) in Figure 5. This type of initialization would be referred to as geopotential constrained non-linear normal mode initialization.

In practice we have observations of both mass and wind, irregular in space and time and with varying accuracies. Some of the data will be inconsistent with other data, so we cannot expect to fit all data and still be on the slow manifold. However, we might have more faith in some observations than others, so we might desire that the adjusted state (a) be on the slow manifold and (b) fit the good data as well as possible and fit the poor data less exactly.

The operational solution of this problem is very complex, because it lies at the ill-defined interface between analysis (which attempts to fit the data according to its presumed accuracy) and initialization (which attempts to suppress high frequency oscillations).

Formally, however, one can write solutions to the constrained initialization problem using the variational calculus. Daley (1978) developed the variational formalism for the constrained normal mode initialization of the shallow water equations and solved it for some simple cases. The variational problem can be posed as follows.

Consider the case where we have observations of the vector wind $\underline{v}$ and geopotential Φ and we indicate our confidence in the observations by means of weight functions $w_{\underline{v}}(\lambda, \phi, P)$, $w_{\Phi}(\lambda, \phi, P)$. Thus, where we had complete confidence in Φ, but none in $\underline{v}$, w_{Φ} would be large and $w_{\underline{v}} = 0$. This would correspond to remaining on a geopotential manifold. Conversely, $w_{\Phi} = 0$, $w_{\underline{v}}$ large would correspond to remaining on a wind manifold. The general solution, with both w_{Φ}, $w_{\underline{v}}$ nonzero would define a manifold somewhere in between the geopotential and wind manifolds. We can pose the general problem of going from model state 0 to model state c in Figure 5 as the minimization of the following functional

$$I = \int_A \left[(\underline{v}_0 - \underline{v}_c)^2 w_{\underline{v}} + (\Phi_0 - \Phi_c)^2 w_{\Phi} \right] dA \qquad (50)$$

where $\int_A [\ ] dA$ is an integral over the atmosphere, $\underline{v}_0, \Phi_0$ indicate the observed values and $\underline{v}_c, \Phi_c$ indicate the values after constrained initialization.

In the variational formulation we wish to minimize (50) subject to the constraint that the final state lie on the slow manifold. We can do this approximately by demanding that the final state satisfy Machenhauer's condition (41) - viz.

$$Z_c = \frac{R_Z(Z_c, Y_c)}{2\Omega i \Lambda_Z} \qquad (51)$$

where Y_c and Z_c are the projection of the Rossby and gravity modes respectively after adjustment has taken place. Constraint (51) can be applied to the functional (50) by means of Lagrange's Undetermined Multipliers. This new functional is then minimized, leading to a set of Euler-Lagrange equations. We note that the projection on the Rossby modes Y will change after a data constrained initialization procedure. Unconstrained initialization is a special case of constrained initialization with a particular choice of w_Φ, $\underline{w_v}$ (see Daley, 1978).

17. NON-LINEAR NORMAL MODE INITIALIZATION AND QUASI-GEOSTROPHIC THEORY

Leith (1980) showed the connection between non-linear normal mode initialization and quasi-geostrophic theory. He confined his attention to the first 2 passes through Machenhauer's scheme (equations 38 and 39) or the Baer-Tribbia scheme (equations 47 and 48). There is a great deal of algebraic manipulation in the derivation, so we will omit all details and concentrate on the results.

The analysis is done on an f-plane with a form of the pressure coordinate equations. Leith (1980) performs the first 2 passes (equations 38 and 39) through Machenhauer's scheme and then converts the results back to real space form. We will consider the case of streamfunction (vorticity) constrained initialization. The results will be given in the pressure coordinate system (x, y, P).

We define $\sigma^* = \frac{R\gamma^*}{P}$ (equation 4) and ψ_I, Φ_I, ω_I to be the initial raw streamfunction, geopotential and vertical motion fields respectively. The Coriolis parameter f is constant. The subscripts 0 and 1 will refer to the results after the first and second steps respectively. As in Section 13, the first step is linear initialization while the second step is a non-linear step.

Streamfunction constrained initialization

Step 1 - $\psi_o = \psi_I$, $\Phi_o = f\psi_o$, $\omega_o = 0$

Step 2 - $\psi_1 = \psi_o = \psi_I$

$$\nabla^2 \Phi_1 = f\nabla^2 \psi_o + 2\left[\frac{\partial^2 \psi_o}{\partial x^2} \frac{\partial^2 \psi_o}{\partial y^2} - \left(\frac{\partial^2 \psi_o}{\partial x \partial y}\right)^2\right] \tag{52}$$

$$\sigma \ast \nabla^2 \omega_1 + f^2 \frac{\partial^2 \omega_1}{\partial p^2} = f \frac{\partial}{\partial p} (\underline{v}_o \cdot \nabla \nabla^2 \psi_o)$$
$$- \nabla^2 (\underline{v}_o \cdot \nabla \frac{\partial \phi_o}{\partial P}).$$
(53)

where $\underline{v}_o = \underline{k} \times \nabla \psi_o$.

We note that step 1 is simply geostrophic initialization, in which the geopotential is calculated from the streamfunction geostrophically. In the second step ϕ and ω are calculated from the streamfunction using the non-linear balance equation and the quasi-geostrophic ω equation respectively. Note that ω_I is not used at all in the calculations.

18. SOME SUCCESSES OF NON-LINEAR NORMAL MODE INITIALIZATION

We shall now demonstrate that non-linear normal mode initialization "works" and that the effort expended is justified. We shall concentrate on the application of the Machenhauer scheme (Section 13) to two baroclinic primitive equation models - the ECMWF model and the Canadian Operational Spectral Model. The results in this section are taken from Temperton and Williamson (1979) and Daley (1979).

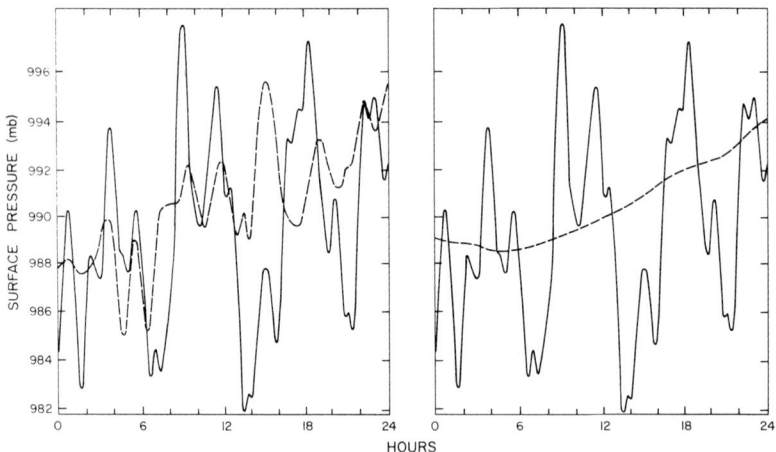

Figure 6 - after Temperton and Williamson (1979)

The acid test of any initialization scheme is - does it suppress high frequency gravity waves? We show in Figure 6 24-hour time traces of surface pressure at a particular gridpoint of the ECMWF model before and after linear and non-linear normal mode initialization. The solid lines (identical in both panels) indicate that without initialization there were high frequency oscillations of more than 10 millibars. The dashed line on the left panel indicates that linear initialization (Section 12) is only partially successful in suppressing gravity waves. The dashed line on the right panel indicates the almost complete success of non-linear normal mode initialization in eliminating the high frequencies. To put this result in perspective, it should be stated that no other initialization scheme is capable of such success when applied to a model of the complexity of the ECMWF model.

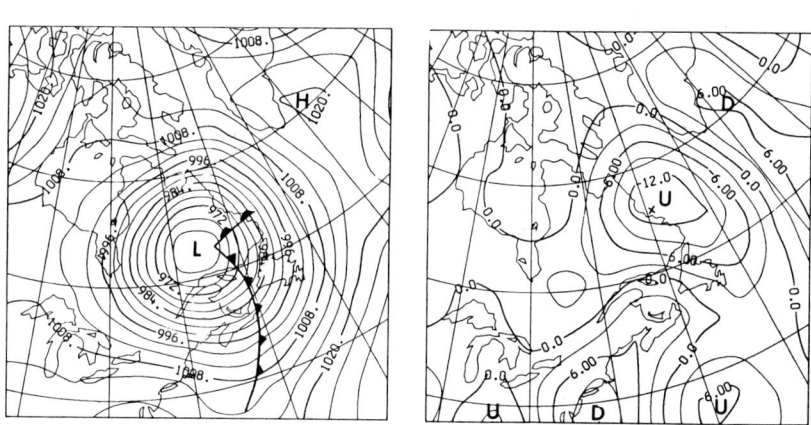

Figure 7 - after Daley (1979)

In addition to their ability to suppress high frequencies, initialization schemes are often required to provide consistent information to make up for deficiencies in the observation network. For example, it is difficult to accurately observe the divergent wind field or vertical motion, and the present initialization scheme has at least the possibility of providing this information. Figure 7 is taken from Daley (1979) and attempts to show the vertical motion field provided by the application of the Machenhauer scheme to the

Canadian Operational Spectral Model. In the left panel is the surface pressure field and in the right panel is the 850 mb vertical motion field produced after non-linear initialization. The contours on the right panel are in mb/hr and U indicates upward motion while D indicates downward motion. It can be seen that the scheme produces upward motion ahead of the warm and cold fronts and downward motion behind the cold front, consistent with synoptic theory.

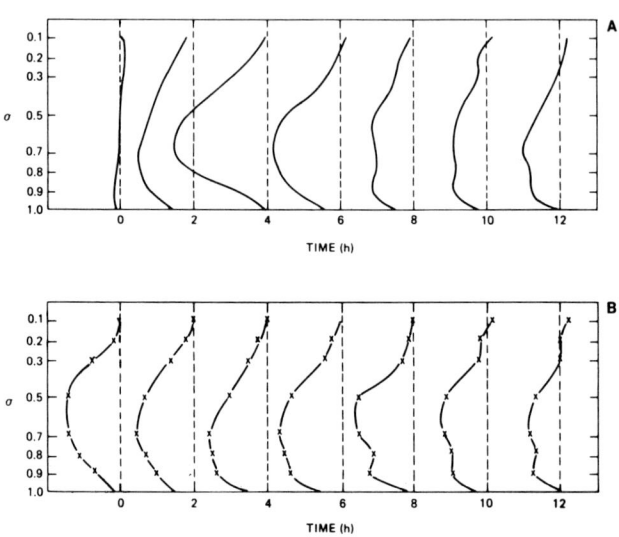

Figure 8 - after Daley (1979)

A second important question concerning the vertical motion field is: does it evolve slowly in time or does it oscillate before settling down? Figure 8 shows the vertical motion profiles as a function of time at the point marked X (Labrador Coast) on the right panel of Figure 7. The upper series is without initialization while the lower is with initialization. It can be seen that without initialization the vertical motion profile goes through an oscillation before settling down, while initialization produces a smoothly evolving sequence of profiles.

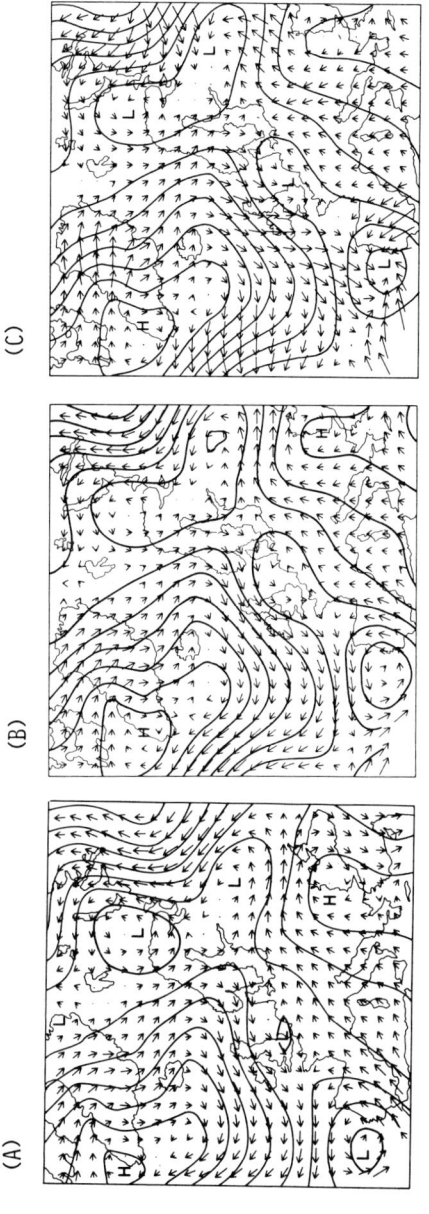

Figure 9 - after Temperton and Williamson (1979)

Non-linear normal mode initialization schemes are also capable of generating cross-isobar flow in the model boundary layer. Figure 9 is taken from Temperton and Williamson (1979). The left panel shows the 1000 mb geopotential height and winds without initialization. Note that the wind flow is more or less geostrophic. In the center panel is shown the same fields after application of the Machenhauer scheme to an <u>adiabatic</u> version of the ECMWF model. The flow has changed, but it is still more or less geostrophic. This is to be expected, because cross isobar flow in the boundary layer is a function of non-adiabatic terms such as the surface stress. The panel on the right shows the flow after application of the Machenhauer method to a version of the ECMWF model containing surface stress terms. Note the cross-isobar flow.

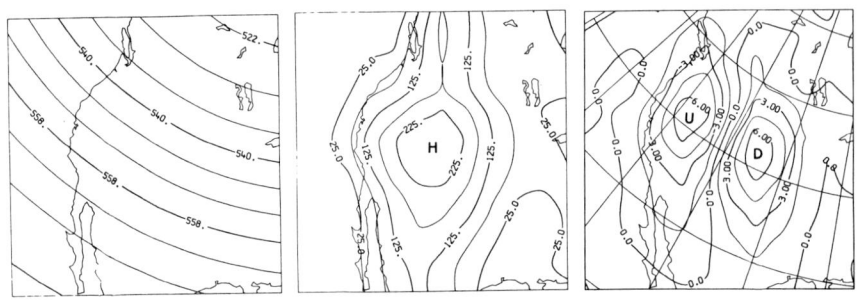

Figure 10 - *after Daley (1979)*

Non-linear normal mode initialization can also generate mountain induced vertical motion. Daley (1979) performed an experiment to demonstrate this. Since the terrain induced vertical motion is smaller in amplitude than the synoptic vertical motion, he started with idealized data in order to suppress the latter. An initially zonally averaged flow in pressure coordinates was interpolated to the terrain following coordinates of the Canadian Operational Spectral Model. Then Machenhauer's scheme was applied. In Figure 10, left hand panel, is shown the 500 mb geopotential height demonstrating the zonal averaging. In the center panel is shown the topography field of the Rocky Mountain Range (dekameters) - the main axis of which is perpendicular to the flow. The right hand panel shows the resulting vertical motion (mb/hr). As hoped, the upstream vertical motion is upward (U) and the downstream vertical motion is downward (D).

19. RECENT DEVELOPMENTS AND RELATED RESEARCH

This review has concentrated on the use of model normal mode procedures in the initialization of primitive equation models. Recent work, however, suggests that the normal mode procedures have considerably wider application in numerical weather prediction. We will briefly discuss several of these more recent applications.

Daley and Puri (1980) have used normal mode procedures in an examination of the four-dimensional data assimilation problem. The model normal modes were found to be particularly useful in diagnosing and partially ameliorating the data rejection problem.

Williamson, Daley and Schlatter* have examined optimal interpolation procedures using the model normal modes. They have shown that the geostrophic constraint used in multi-variate optimal interpolation procedures can both improve and degrade the analysis quality under certain conditions. Some experiments have been made with more realistic and general constraints than geostrophic.

Daley, Tribbia and Williamson* have examined the spurious excitation of large-scale external Rossby waves using the normal modes of barotropic and baroclinic models. In particular, it has been found that poor analysis methods in the tropics, or the imposition of an equatorial wall can excite spuriously large free Rossby modes. An attempt to initialize these modes using Machenhauer's technique was successful.

Normal mode methods are being used in more and more of the world's meteorological centers. We expect that these techniques will be used to gain understanding of the complex primitive equation models in use today and hopefully lead toward improvements in their performance.

---

* Paper in preparation.

References

Andersen, J., 1977: A routine for normal mode initialization with non-linear correction for a multi-level spectral model with triangular truncation. ECMWF Internal Report No. 15, 41 pp. (Available from ECMWF library)

Baer, F., 1977: Adjustments of initial conditions required to suppress gravity oscillations in non-linear flows. Beitrage zur Physik der Atmosphäre, 50, 350-366.

Baer, F. and Tribbia, J., 1977: On complete filtering of gravity modes through non-linear initialization. Mon. Wea. Rev., 105, 1536-1539.

Ballish, B., 1979: Comparison of some non-linear initialization techniques. Preprint volume - Fourth Conference on Numerical Weather Prediction, Silver Spring, Maryland, October, 1979. 9-12.

Blumen, W., 1972: Geostrophic adjustment. Reviews of Geophysics and Space Physics, 10, 485-528.

Daley, R., 1978: Variational non-linear normal mode initialization. Tellus, 30, 201-218.

Daley, R., 1979: The application of non-linear normal mode initialization to an operational forecast model. Atmosphere-Ocean, 17, 97-124.

Daley, R., 1980: On the optimal specification of the initial state for deterministic forecasting. Accepted for publication in Mon. Wea. Rev.

Daley, R., 1981: Normal mode initialization. Rev. Geophys. Space Phys., 19, (in press).

Daley, R. and Puri, K., 1980: Four-dimensional data assimilation and the slow manifold. Mon. Wea. Rev., 108, 85-99.

Flattery, T., 1967: Hough functions. Technical Report No. 21, The University of Chicago, Department of the Geophysical Sciences.

Flattery, T., 1970: Spectral models for global analysis and forecasting. Proc. Sixth AWS Technical Exchange Conference, U.S. Naval Academy, Air Weather Service Tech. Report 242, 42-53.

Hough, S., 1898: On the application of harmonic analysis to the dynamical theory of the tides - Part II. On the general integration of Laplace's dynamical equations. Phil. Trans. Roy. Soc. London, A191, 139-185.

Kasahara, A., 1976: Normal modes of ultra-long waves in the atmosphere. Mon. Wea. Rev., 104, 669-690.

Kasahara, A., 1978: Further studies on a spectral model of the global barotropic primitive equations with Hough harmonic expansions. J. Atmos. Sci., 35, 2043-2051.

Kasahara, A., 1980: Effect of zonal flows on the free oscillations of a barotropic atmosphere. J. Atmos. Sci., 37, 917-929.

Kasahara, A. and Puri, K., 1980: Spectral representation of three-dimensional global data by expansion in normal mode functions. NCAR manuscript 0501/79-8. Available from NCAR, Boulder, Colorado 80307, U.S.A.

Leith, C., 1980: Non-linear normal mode initialization and quasi-geostrophic theory. J. Atmos. Sci., 37, 958-968.

Longuet-Higgins, M., 1968: The eigenfunctions of Laplace's tidal equations over a sphere. Phil. Trans. Roy. Soc. London, A262, 511-607.

Machenhauer, B., 1977: On the dynamics of gravity oscillations in a shallow water model, with applications to non-linear normal mode initialization. Beitrage zur Physik der Atmosphäre, 50, 253-271.

Taylor, G., 1936: The oscillations of the atmosphere. Proc. Roy. Soc. London, A156, 318-326.

Temperton, C. and Williamson, D., 1979: Normal mode initialization for a multi-level gridpoint model. ECMWF Technical Report No. 11, 99 pp. (Available from ECMWF library)

Tribbia, J., 1979: Non-linear initialization on an equatorial beta plane. Mon. Wea. Rev., 107, 704-713.

Tribbia, J., 1980: Normal mode balancing and the ellipticity condition. Submitted to Mon. Wea. Rev.

Tribbia, J., 1980: A note on variational non-linear normal mode initialization. Submitted to Mon. Wea. Rev.

Williamson, D., 1976: Normal mode initialization procedure applied to forecasts with the global shallow water equations. Mon. Wea. Rev., 104, 195-206.

Williamson, D., and Dickinson, R., 1976: Free oscillations of the NCAR global circulation model. Mon. Wea. Rev., 104, 1372-1391.

ASSIMILATION OF ASYNOPTIC DATA
AND
THE INITIALIZATION PROBLEM

K. P. Bube[*] and M. Ghil[**]
Courant Institute of Mathematical Sciences
New York University, New York, N. Y. 10012

ABSTRACT

We discuss a mathematical framework for the use of asynoptic data in determining initial states for numerical weather prediction (NWP) models. A set of measured data, synoptic and asynoptic, is termed complete if it determines the solution of an NWP model uniquely. We derive theoretical criteria for the completeness of data sets. The practical construction of the solution from a complete data set by intermittent updating is analyzed, and the rate of convergence of some updating procedures is given.

It is shown that the time history of the mass field constitutes a complete data set for the shallow-water equations. Given that the time derivatives of the mass field are small at initial time, we prove that the velocity field obtained by the diagnostic equations we derive will also have small time derivatives. Hence our diagnostic equations also solve the initialization problem for this system, namely they provide an initial state which leads to a slowly evolving solution to the system.

Finally, we review the bounded derivative principle of Kreiss. It states that in systems with a fast and a slow time scale, initial data can be chosen so that the solution starts out slowly. For such initial data, the solution will actually stay slow for a length of time comparable to the slow time scale. The application of the principle to the initialization problem of NWP is discussed.

---
[*] Also Department of Mathematics, University of California, Los Angeles, CA 90024

[**] Also Laboratory for Atmospheric Sciences, NASA Goddard Space Flight Center, Greenbelt, MD 20771

1. INTRODUCTION

Numerical weather prediction (NWP) is an initial-value problem for a system of nonlinear partial differential equations in which the initial values are known only incompletely and inaccurately. Data at initial time can be supplemented, however, by observations of the system distributed over a time interval preceding it.

A large number of observations is made by the conventional, ground-based meteorological network. They consist of point values of temperature, humidity, pressure and horizontal velocity. These observations are produced at the so-called synoptic times, 0000 GMT and 1200 GMT. It is customary therefore to choose a synoptic time as initial time for a numerical forecast. Conventional observations are insufficient in number in order to determine the initial state of the model atmosphere. Furthermore, they are very unevenly distributed in space, being concentrated over the continents of the Northern Hemisphere, and much sparser over the oceans and over the Southern Hemisphere.

A large number of additional observations are made at the so-called subsynoptic times, 0600 GMT and 1800 GMT, as well as in an essentially time-continuous manner, using geostationary satellites, polar-orbiting satellites and other nonconventional measuring platforms. All these observations together are called asynoptic.

The object of four-dimensional (4-D) data assimilation is to construct a set of complete, accurate initial data for a NWP model from the measured data, synoptic as well as asynoptic. The completeness question for the set of measured data can be formulated as follows: do these measured data, with their distribution over a time interval and over space, determine the solution of the model equations uniquely? Furthermore, do they determine this solution in a way which depends continuously on the data, so that small errors in the data only cause small errors in the solution?

To understand better the accuracy question, we have to recall that the primitive equations, which govern most operational NWP models today, admit two types of solutions: slow, quasi-geostrophic, meteorologically significant motions, and fast inertia-gravity waves. The latter are present in the real atmosphere only with very small amplitudes. In a model they interfere with short-range forecasts, up to 12h, of all fields of motion, and with longer-range forecasts of vertical velocity and precipitation. It is desirable therefore to eliminate entirely these waves from the model at initial time, or to reduce them as much as possible. To achieve this elimination or reduction is the purpose of initialization.

The goal of preparing initial data in NWP can be restated thus as that of going from a set of inaccurate data, both synoptic and asynoptic, to a complete set of synoptic data which will not generate fast waves when starting a forecast from it. Until rather recently, the meteorological literature has handled separately the two problems of 4-D assimilation, i.e., of passing from nonstandard, combined synoptic and asynoptic data to complete synoptic data, and that of initialization, i.e., of modifying synoptic data to prevent the growth of fast waves in the ensuing forecast.

The purpose of this review is to outline a number of approaches which have contributed to put both aspects of data handling in NWP on a more solid mathematical foundation, as well as pointing in the direction of their eventual unification. A companion article in this volume (Ghil et al., 1980) attempts to provide a systematic, unified theory of 4-D data assimilation and initialization.

In the next section, we shall deal mathematically with the problem of assimilating nonstandard data. The most common practical procedure to use data available at times preceding initial forecast time is intermittent updating. It was suggested by Charney et al. (1969) and independently by Smagorinsky et al. (1970). The model is provided the best available data at some preceding instant, e.g., 24 h or 48 h earlier, and is integrated forward in time. Additional data replace the model values as they become available: the model is updated. When the model integration reaches the initial instant for the next scheduled forecast, its guess of the initial state is blended with the data available at that instant to produce the desired estimate. Thus the model itself is used to assimilate the data available up to the initial instant. Variations of this procedure, as well as other procedures for the four-dimensional (4-D), space-time assimilation of meteorological observations are reviewed in Bengtsson (1975).

Our discussion in Section 2 will be based on the work of Bube (1978, 1980). Connections between the theoretical completeness question and the practice of intermittent updating will be pointed out.

In Section 3, another approach to completeness, based on the work of Ghil (1975; Ghil et al., 1977) will be discussed.

Connections with the initialization problem (Ghil, 1980) will also be outlined.

In Section 4, one particular aspect of the initialization question will be addressed. Various procedures for eliminating or reducing inertia-gravity waves exist, such as the nonlinear normal-mode approach of Baer and Tribbia (1977) or of Machenhauer (1977), and they are reviewed in this volume by Daley (1980). After applying any such procedure, we still have to ask: for how long after initial time has the rapid growth of the fast waves been prevented?

This question is answered by the bounded derivative principle of Kreiss (1979, 1980). Given that the time derivatives of the solution are small initially, these derivatives will stay small for a length of time comparable to the system's slow time scale, 6-24 h say, rather than merely to its fast time scale, O(10 min). The application of the principle to a system of equations of interest in NWP is illustrated. An initialization procedure based on the principle's formalism (Browning et al., 1980) is sketched and its connection with the completeness question and with other initialization procedures is commented upon.

Concluding remarks follow in Section 5.

2. COMPLETENESS AND UPDATING

In this section we review some mathematical results which address the completeness question for the shallow-water equations. These equations have solutions which share some of the essential properties of large-scale atmospheric flow. The initial-value problem for this system of partial differential equations is the problem of determining a solution of the system from complete initial data, i.e., from the global specification of all the state variables at a single time. For this system, the initial-value problem is well posed: given the initial data, there exists a unique solution of the system, and this solution depends continuously on the initial data.

We call nonstandard any set of measured data other than a complete set of initial data. The completeness question for a nonstandard data set can be formulated as follows: is it possible to determine complete initial data at some time from the nonstandard data set, with the initial data depending continuously on the measured nonstandard data set? There are really two separate questions to be answered -- the theoretical uniqueness and stability question for the differential equations, and the practical question of computationally constructing initial data from measurements of a nonstandard data set.

For the sake of brevity, we consider only the case where the nonstandard data set consists mainly of measurements of the mass fields (i.e., temperature and surface pressure), with few measurements of the wind fields. This case is of some historical interest because the large number of mass field measurements available from polar-orbiting satellites was the first important set of nonconventional data used in NWP.

We shall consider in the sequel various forms of the shallow-water equations, in order to illustrate as directly as possible each one of a number of theoretical questions we wish to address. The full, nonlinear system can be written in cartesian coordinates on a plane tangent to the Earth at a latitude θ_0, say, as:

$$u_t + uu_x + vu_y = -\phi_x + fv, \qquad (2.1a)$$

$$v_t + uv_x + vv_y = -\phi_y - fu, \qquad (2.1b)$$

$$\phi_t + u\phi_x + v\phi_y = -\phi(u_x + v_y). \qquad (2.1c)$$

Here x is the coordinate pointing in the zonal, West-East direction, y in the meridional, South-North direction, with u and v the corresponding velocity components. The height of the free surface is h, with $\phi = gh$ the geopotential, g being the acceleration of gravity. The Coriolis parameter f is taken to be constant, $f = 2\Omega \sin \theta_0$, Ω being the angular velocity of the Earth.

We shall be also interested in a linearized form of the equations; the linearization is made around a state with $u = U$, $v = 0$, $\phi = \Phi$, which satisfies the geostrophic constraint, $fU = -\Phi_y =$ const. (see also Ghil et al., 1980). The linear system is

$$u_t + Uu_x = -\phi_x + fv, \qquad (2.2a)$$

$$v_t + Uv_x = -\phi_y - fu, \qquad (2.2b)$$

$$\phi_t + U\phi_x = -\Phi(u_x + v_y) + fUv. \qquad (2.2c)$$

2.1 The completeness question

Bube and Oliger (1977) and Bube (1978, 1980) have considered completeness questions generally for linear first-order hyperbolic systems where the nonstandard data set consists essentially of measurements of a fixed set of the state variables, such as the mass field. We present here the results for the model system

$$\phi_t = a\phi_x + bu_x, \qquad (2.3a)$$

$$u_t = b\phi_x + au_x, \qquad (2.3b)$$

with periodic boundary conditions in x. The linearized shallow-water equations (2.2) for purely zonal, $v \equiv 0$, one-dimensional flow around the circle of latitude $\theta = \theta_0$ can be written in this form, where ϕ is the geopotential, u is the zonal velocity, and the constants $-a$ and $-b$ are U and $\sqrt{\Phi}$, the zonal velocity and the geopotential of the basic state, respectively.

We consider the theoretical question first. We want to find nonstandard data sets, consisting as much as possible of measurements of ϕ only, from which complete initial data, say at time $t = 0$, can be determined. Two important observations can be made immediately. First, in order to be able to infer anything about u from measurements of ϕ, the equations must have sufficient linkage between ϕ and u; for system (2.3), we must have $b \neq 0$. For the linearized shallow-water equations, this is satisfied since $\Phi > 0$. Second, u can be determined at most up to a constant from measurements of ϕ alone, since $\phi = 0$, $u = u_0$ is a solution of system (2.3). It follows that at least one measurement of u is needed.

The periodic domain in x will be the unit interval $0 \leq x \leq 1$. We examine three nonstandard sets of data:
(i) measurements of $\phi(x,0)$ and $\phi_t(x,0)$ for $0 \leq x \leq 1$, and of $u(0,0)$;
(ii) measurements of $\phi(x,t)$ for $0 \leq x \leq 1$ and $t = -j\tau$ for some $\tau > 0$ and $j = 0,1,\ldots,m$, and of $\int_0^1 u(x,0)\,dx$;
(iii) measurements of $\phi(x,t)$ for $0 \leq x \leq 1$ and $-T \leq t \leq 0$ for some $T > 0$, and of $\int_0^1 u(x,0)\,dx$.

For data set (i), the "instantaneous time-history" of ϕ at time $t = 0$ is measured, i.e., its time derivative, in addition to $\phi(x,0)$.

Assuming $b \neq 0$, Eq. (2.2a) becomes an ordinary differential equation for $u(x,0)$, which we can solve using the measured value of $u(0,0)$. For the solution $u(x,0)$ to be periodic, the measured $\phi_t(x,0)$ must satisfy the compatibility condition

$$\int_0^1 \phi_t(x,0)\, dx = 0. \tag{2.4}$$

Thus data set (i) is a good nonstandard data set: we can determine from it $u(x,0)$, and $u(x,0)$ depends continuously on the measured data. This approach corresponds to the use of a generalized set of diagnostic equations to complete a set of initial data. It will be discussed further in the next section.

For data set (ii), we sample the time history of ϕ at discrete, periodic points in time, with our last measurement at the initial time for a forecast. This measurement pattern is closer than (i) to the way data are gathered operationally; in fact, data are measured on even more complicated space-time manifolds. To analyze the completeness question for data sets (ii) and (iii), we expand ϕ and u in Fourier series in x:

$$\phi(x,t) = \sum_{\xi=-\infty}^{\infty} \hat{\phi}(\xi,t)\, e^{2\pi i \xi x}, \tag{2.5a}$$

$$\hat{\phi}(\xi,t) = \int_0^1 e^{-2\pi i \xi x}\, \phi(x,t)\, dt, \tag{2.5b}$$

where ξ is the wave number. Note that

$$\hat{u}(0,0) = \int_0^1 u(x,0)\, dx. \tag{2.6}$$

It can be shown that if τb is a rational number, then data set (ii) does not determine $u(x,0)$ uniquely. If τb is irrational, then data set (ii) determines $u(x,0)$ uniquely, but $u(x,0)$ does not depend continuously in the L^2-norm on the measured data, i.e., root-mean-square (rms) errors in $u(x,0)$ will not be small if the rms errors in the measured data are small.

For data set (iii), however, we do have continuous dependence. In fact, for each $T > 0$, there is a constant C_T for which

$$\| u(x,0) \| \leq C_T (\hat{u}(0,0) + \max_{-T \leq t \leq 0} \| \phi(x,t) \|), \qquad (2.7)$$

where $\| \ \|$ is the L^2 or rms norm,

$$\| y(x) \|^2 = \int_0^1 |y(x)|^2 \, dx. \qquad (2.8)$$

Estimate (2.7) immediately implies both uniqueness and continuous dependence.

2.2 Computational construction of initial data.

We now address the second question, of actually constructing the initial data from nonstandard, measured data. In particular, we consider the method of <u>intermittent updating</u>, where the measured values of ϕ are inserted into a computation as the computation proceeds. At first it appears that our results on data set (ii) imply that intermittent updating does not use enough data to determine $u(x,0)$ even theoretically in a continuous manner. In practice, however, only a finite set of measurements can be used for any computation, and we cannot hope to determine a function of $\underline{x}$ completely for all x in the interval $0 \leq x \leq 1$ either. If our discrete measurements of $\phi(x,t)$ are sufficiently dense in x and t, then these measured data are a good approximation to $\phi(x,t)$ in the norm $\max_{-T \leq t \leq 0} \| \phi(x,t) \|$, and data set (iii) is actually the appropriate theoretical model for the measured data set. Thus, theoretically, enough data are being used. We now examine how intermittent updating uses this data set.

For this purpose, we shift the time origin back to the beginning of a numerical computation. Let τ be the updating interval; we have measurements of $\phi(x,t)$ for $t = j\tau$, $j = 0,1,2,\ldots$. We start with the initial measurement of ϕ and an initial approximation to u at $t = 0$. As we integrate forward in time, we replace the computed values of ϕ by the measured values at times $t = j\tau$.

Let $\varepsilon(x,t)$ denote the error in the computed u at time t. Notice that ε is not reduced <u>at</u> the updating times, since u is <u>not</u> updated:

only ϕ is. The error in u will decrease between updating times, as a result of the linkage in the system beween u and ϕ.

Expanding ε in its Fourier series and ignoring the distinction between the numerical solution and the corresponding solution of the differential equation, we can show that

$$\hat{\varepsilon}(\xi,(j+1)\tau) = \rho(\xi,\tau)\;\hat{\varepsilon}(\xi,j\tau), \qquad (2.9)$$

where

$$\rho(\xi,\tau) = |\cos(2\pi b\xi\tau)|. \qquad (2.10)$$

So the effect of an update is to multiply each Fourier coefficient of the error in u by a decrease factor $\rho(\xi,\tau) \leq 1$. If $\rho(\xi,\tau) \ll 1$, then $\hat{\varepsilon}(\xi,j\tau)$ approaches 0 rapidly as j increases. If $\rho(\xi,\tau) \approx 1$, then $\hat{\varepsilon}(\xi,j\tau)$ approaches 0 slowly as j increases.

For $t = j\tau$,

$$|\hat{\varepsilon}(\xi,t)| = \sigma(\xi,\tau)^t |\hat{\varepsilon}(\xi,0)|, \qquad (2.11)$$

where

$$\sigma(\xi,\tau) = \rho(\xi,\tau)^{1/\tau}. \qquad (2.12)$$

It can be shown that, for $\xi \neq 0$, $\sigma(\xi,\tau)$ is a strictly decreasing function of τ as long as $0 < \tau < 1/(4|b\xi|)$, and

$$\lim_{\tau \to 0} \sigma(\xi,\tau) = 1. \qquad (2.13)$$

In other words, as τ decreases, σ increases monotonically and tends to 1. Hence updating ϕ more frequently does not necessarily make $\varepsilon(x,t) \to 0$ faster as t increases. We must allow enough time for the new information to pass from ϕ to u; alternatively, the energy of the error has to have time to pass from u to ϕ, and then out of the system when ϕ is updated.

Note the important dependence of ρ and σ on both wave number ξ and updating interval τ. First, if $\xi = 0$, then $\rho = \sigma = 1$, so no improvement is made in the mean velocity $\hat{u}(0,t)$. If $\varepsilon(x,t)$ is to converge to 0, the value of $\hat{u}(0,0)$ in the initial approximation to u must be correct. For other values of ξ, the best updating interval is

$$\tau \approx \frac{1}{4|b\xi|} .$$

This result for τ has an intuitive explanation. Consider initial data for system (2.3) of the form

$$\phi = 0 , \qquad v = v_0 \sin(2\pi\xi x) .$$

In other words, the error in geopotential is zero, while the error in velocity is purely in wave number ξ. The corresponding solution of (2.3) has

$$\phi(x,t) = v_0 \sin(2\pi\xi bt) \cos[2\pi\xi(at+x)] .$$

The geopotential error at time t generated by the velocity error at time $t = 0$ is thus also a pure ξ-wave, travelling at phase speed $\underline{a}$, but amplitude-modulated with frequency $\omega = |b\xi|$. Hence our result means that the best updating interval for this wave of frequency ω is $\tau \approx 1/4\omega$, i.e., 1/4 of the period of this amplitude modulation.

We are interested in what happens if the frequency of updates were increased. For fixed $\xi \neq 0$ and "assimilation cycle length" $t > 0$, letting the update interval τ decrease, $\tau \to 0$, Eqs. (2.11, 2.13) imply that

$$|\hat{\varepsilon}(\xi,t)| \to |\hat{\varepsilon}(\xi,0)| ,$$

i.e., the error at time t is not decreased from that at $t = 0$, even though ϕ was updated more often and more data were used. In other words, $\tau \approx 1/4\omega$ is really an optimum, and further decrease of τ is counterproductive.

If the entire error is in only one wave number ξ, i.e., if $\hat{\varepsilon}(\xi,0) \neq 0$ only for one value of ξ, then τ can be chosen judiciously to decrease the error quickly. In practice, $\hat{\varepsilon}(\xi,0) \neq 0$ for many values of ξ. The trouble is that if $\tau \gg 1/(4|b\xi|)$, then $\rho(\xi,\tau)$ is just as likely to be close to 1 as it is to be close to 0. The best τ for a small value of $|\xi|$ may give a $\rho \approx 1$ for some large values of $|\xi|$. The dilemma becomes: if τ is too large, waves with high wave number may converge slowly; if τ is too small, waves with low wave number will converge slowly.

One possible solution to this dilemma is to use several different updating intervals $\tau_1,\ldots,\tau_m$, and repeat them in a cycle. If there is a $\tau_j \approx 1/4\omega$ for each frequency ω of interest, then convergence should be reasonably rapid for all corresponding wave numbers. Numerical experiments with model system (2.3) confirm all the theoretical results above.

This analysis extends directly to the case of forward and backward integration in an update cycle: one such cycle is equivalent to two updating intervals. Talagrand (1980) has also analyzed forward-backward updating schemes.

3. COMPLETENESS AND INITIALIZATION

In the previous section we have seen that, for the one-dimensional, linear system (2.3), instantaneous mass field data $\phi(x,0)$ and $\phi_t(x,0)$ determine the velocity field $u(x,0)$, provided the velocity at a single point, $u(0,0)$ say, is given and that (2.4) holds. In this section we shall show first that similar results hold for a two-dimensional linear version of the shallow-water equations, as well as for the full, nonlinear system (2.1). Then it will be shown that complete initial data determined by the procedure outlined here from nonstandard mass field data lead to a solution of (2.1) in which fast waves have moderate amplitude.

3.1 Diagnostic equations for compatible balancing

We consider first the shallow-water equations linearized around a state of rest, $U = 0$, $\Phi = $ const.:

$$u_t = -\phi_x + fv, \qquad (3.1a)$$

$$v_t = -\psi_y - fu, \qquad (3.1b)$$

$$\phi_t = -\phi(u_x + v_y) \, . \qquad (3.1c)$$

We deal here only with this special case of (2.2), in which U = 0, for the sake of simplicity.

Our purpose is to derive a set of equations for u and v at time t = 0, say, given ϕ and its "instantaneous time history" ϕ_t, as well as higher time derivatives, if necessary, at t = 0.

Clearly (3.1c) is such a <u>diagnostic</u> equation for u,v, given ϕ_t. In the present, two-dimensional case, this equation does not determine u and v completely, while (2.3a) did determine u completely for the previous, one-dimensional case.

A second diagnostic equation can be derived easily. Differentiate (3.1a) with respect to x, (3.1b) with respect to y and (3.1c) with respect to t. The quantities u_{xt} and v_{yt} appearing in the differentiation of (3.1c) can then be substituted from their values obtained when differentiating (3.1a) and (3.1b), respectively. This leads to the diagnostic system (Ghil, 1975, 1980):

$$u_x + v_y = -\phi_t/\phi, \qquad (3.2a)$$

$$u_y - v_x = -(1/f)(\nabla^2\phi - \phi_{tt}/\phi), \qquad (3.2b)$$

where $\nabla^2\phi = \phi_{xx} + \phi_{yy}$. Thus ϕ_{tt}, as well as ϕ_t, is necessary here to determine u and v at t = 0.

The linear system (3.2) is a set of inhomogeneous Cauchy-Riemann equations for the functions v, u; cross-differentiation of these equations would lead to a Poisson equation for either u or v. A boundary-value problem having a unique, stable solution for (3.2) is the Dirichlet problem: prescribing u, say, on a closed contour ∂D. These boundary conditions determine u completely and v up to an additive constant, in the domain D bounded by the contour ∂D; the value of this constant can be given by prescribing v at some point in D. Alternatively, one can prescribe v on ∂D and u at one point. Also u can be given on one part of ∂D and v on the rest. Either one of these possibilities corresponds in the present case to the prescription of u(0,0) in the previous section. Appropriate compatibility conditions, similar to (2.4), are discussed in Ghil and Balgovind (1979).

Hence, within the framework of model (3.1), solving the Dirichlet problem for system (3.2) solves the <u>completeness</u> problem for the nonstandard data set comprised of ϕ, ϕ_t and ϕ_{tt} measured in D, and of u and v measured in a very small subset of D. Furthermore, this solution is stable, i.e., it depends continuously on the nonstandard data.

We are ready to turn our attention to the full, nonlinear system (2.1). A diagnostic system for u and v at t = 0, given ϕ, ϕ_t and ϕ_{tt} at t = 0 can also be derived (Ghil, 1975; Ghil et al., 1977). The procedure is similar to that leading from (3.1) to (3.2), with the material derivative $d/dt = \partial/\partial t + u\,\partial/\partial x + v\,\partial/\partial y$ replacing the partial time derivative $\partial/\partial t$ in the cross-differentiation. The resulting diagnostic system is:

$$u_x + v_y = -\Psi_x u - \Psi_y v - \Psi_t, \qquad (3.3a)$$

$$u_x^2 + 2u_y v_x + v_y^2 + f(u_y - v_x)$$

$$= f(\Psi_x v - \Psi_y u) + \frac{d\Psi_t}{dt} + u\frac{d\Psi_x}{dt} + v\frac{d\Psi_y}{dt}$$

$$-\frac{1}{\phi}(\phi_x^2 + \phi_y^2) - \nabla^2 \phi, \qquad (3.3b)$$

where $\Psi = \log \phi$. More general diagnostic systems for the wind field, given instantaneous mass field history information, have been derived in Ghil (1975) for three-dimensional primitive-equation systems closely related to those used in operational NWP.

System (3.3) reduces, when $\delta = 0 = d\delta/dt$, with $\delta = u_x + v_y$ being the divergence of the wind field, to the classical Monge-Ampère equation. Its solution, which raises interesting mathematical, numerical and meteorological questions was studied in Ghil et al. (1977). Our system, which generalizes this equation, seems to have a unique, stable solution in most situations of meteorological importance. This settles for our purposes the completeness question with respect to system (2.1) and the nonstandard data set comprised of measurements of ϕ, ϕ_t and ϕ_{tt} at one time instant.

Given measurements of ϕ at discrete time intervals t_j, $\phi_j(x,y) = \phi(x,y,t_j)$, this approach would appear to suggest the possibility of constructing initial data in a way different from updating (cf. also Sec. 2.2, first paragraph). Namely, the history of the mass field, $\phi(x,y,t)$, could be interpolated from $\phi_j(x,y)$; furthermore, the partial time derivatives ϕ_t and ϕ_{tt} can be computed at the last measurement time, t_m ($m \geq 2$), which is the initial forecast time. Then u and v could be computed at $t = t_m$ from system (3.3).

In fact, the nonstandard data sets available today are much more complex, and the completeness question from mass field data is of interest mostly from the historical viewpoint and as a relatively simple illustration. From this illustrative perspective, we shall proceed to point out the connection with the initialization question. Specifically, we wish to show that a solution of the prognostic system (3.1), given initial data satisfying (3.2), will contain no fast waves of large amplitude, provided the mass field data had small ϕ_t and ϕ_{tt} at $t = 0$. A similar result will be outlined for (2.1) and (3.3).

3.2 Compatible balancing and initialization

Various systems of diagnostic equations have been considered in the past under the name of balance equations (Bengtsson, 1975, Ch. 6). The idea of static balancing is precisely that of obtaining a complete initial state which leads to a balanced, i.e., quasi-geostrophic, slowly evolving solution to the prognostic equations. We shall show that the solution to system (3.2), which is dynamically compatible with (3.1), does indeed produce such a slow evolution of the solution to (3.1), when it is used as its initial state.

Define the quantities

$$G \equiv \phi_x - fv, \qquad (3.4a)$$

$$H \equiv \phi_y + fu. \qquad (3.4b)$$

It follows from (3.4) and (3.2) that G and H satisfy (Ghil, 1980):

$$G_x + H_y = \phi_{tt}/\Phi, \qquad (3.5a)$$

$$G_y - H_x = (f/\Phi)\phi_t. \qquad (3.5b)$$

Eqs. (3.5) are a system with the same structure as (3.2). Under suitable boundary conditions, its solution (G,H) depends continuously on the right-hand side (ϕ_{tt}, ϕ_t). That is, smallness of ϕ_t and ϕ_{tt} will imply smallness of G and H.

But (3.1a,b) state that $u_t = G$, $v_t = H$. Hence, provided ϕ_t and ϕ_{tt} are small at t = 0, u_t and v_t for a solution of (3.1) obtained from an initial state given by (3.2) will also be small at t = 0. For a discussion of cases in which ϕ_t and ϕ_{tt} are not small at t = 0, see Ghil (1980). We conclude that the diagnostic system (3.2) solves the initialization problem for (3.1), given a set of nonstandard data ϕ, ϕ_t and ϕ_{tt} at t = 0. It can be shown further that small ϕ_{ttt} will lead to small u_{tt} and v_{tt}, and so on: higher temporal smoothness of the instantaneous mass field data will lead to higher smoothness of the initial velocity tendencies.

For the nonlinear shallow-water equations (2.1) and the corresponding, dynamically compatible diagnostic system (3.3), the proof is similar. It is technically more complex, and will only be sketched here (cf. Ghil, 1980).

Let the variables be nondimensionalized by

$$x = Lx', \quad y = Ly', \quad u = Uu', \quad v = Uv', \tag{3.6a}$$

and

$$\phi = \Phi + \phi_0 \phi', \tag{3.6b}$$

with Φ the mean geopotential and ϕ_0,

$$\phi_0 = LfU,$$

a characteristic magnitude of the geopotential's deviation from Φ. Introduce also the Rossby radius of deformation

$$\lambda = \sqrt{\bar{\Phi}}/f \tag{3.7}$$

and the nondimensional Rossby number

$$\varepsilon = U/Lf, \tag{3.8}$$

which is small for midlatitude large-scale flow. We shall assume that (A0):

$$L/\lambda = O(1), \tag{3.9}$$

which in particular means that inertia-gravity waves have velocities roughly equal to pure gravity waves. This permits us to introduce two nondimensional time scales, a <u>fast time</u> t^* and a <u>slow time</u> τ, by

$$t^* = ft, \tag{3.10a}$$

$$\tau = (U/L)t = \varepsilon t^*. \tag{3.10b}$$

For brevity, we shall use the symbol ∂_t for

$$\partial_t = \partial/\partial t^* + \varepsilon\, \partial/\partial\tau, \tag{3.11}$$

and drop primes in the sequel.

We make now the two assumptions that: (A1) the geopotential data satisfy

$$\phi_x, \phi_y = O(1), \tag{3.12a}$$

$$\partial_t \phi, \partial_t^2 \phi = O(\varepsilon); \tag{3.12b}$$

and that (A2) the solution (u,v) of (3.3) with such data has a characteristic length $L_{(u,v)}$ which equals that of the data, $L_{(\phi)}$,

$$L_{(u,v)} = L_{(\phi)} = L, \tag{3.13a}$$

and a characteristic magnitude U such that the corresponding Rossby number R_0 is indeed small

$$R_0 \equiv U/fL = \varepsilon \ll 1 \ . \tag{3.13b}$$

Assumption (A2) actually follows from (A1) in certain cases, and seems to be plausible in general.

Defining G and H as before, in nondimensional form though,

$$G = \phi_x - v \ , \tag{3.14a}$$

$$H = \phi_y + u \ , \tag{3.14b}$$

yields immediately

$$G_x + H_y = \nabla^2 \phi + u_y - v_x \ , \tag{3.15a}$$

$$G_y - H_x = -(u_x + v_y) \ . \tag{3.15b}$$

The right-hand sides of (3.15), however, cannot be equated directly with ϕ_{tt} and ϕ_t, respectively, as was the case for the linear system. Instead, system (3.3) has to be rewritten in nondimensional form, and Eqs. (3.7-3.13) used to show that in fact

$$G_x + H_y = \{1 + O(1)\} \, O(\varepsilon) \ , \tag{3.16a}$$

$$G_y - H_x = O(\varepsilon) \ . \tag{3.16b}$$

From (3.16) we can then conclude, as for (3.5), that $G, H = O(\varepsilon)$. Using (2.1) and (3.14), it follows that

$$du/dt = \partial_t u + \varepsilon(u \, \partial u/\partial x + v \, \partial u/\partial y) = O(\varepsilon) \ ,$$

$$dv/dt = \partial_t v + \varepsilon(u \, \partial v/\partial x + v \, \partial v/\partial y) = O(\varepsilon) \ .$$

This in turn implies $\partial_t u = O(\varepsilon)$, $\partial_t v = O(\varepsilon)$ at $t = 0$.

We saw in Sec. 3.1 that the nonstandard data set comprised of the "instantaneous mass field history" $(\phi, \phi_t, \phi_{tt})$ at $t = 0$

determines completely an initial data set (ϕ, u, v) for the nonlinear shallow-water equations (2.1). The wind field is determined from the diagnostic system (3.3). We have shown here that (3.3) is dynamically compatible with the prognostic system (2.1), which is equivalent to the statement that the smallness of ϕ_t and ϕ_{tt} measured at t = 0 will result in small u_t and v_t at t = 0.

This leaves open the question of how long will (ϕ_t, u_t, v_t), after initialization by compatible balancing, as above, or any other correct initialization procedure (Daley, 1980), stay small. This question is addressed in the next section.

4. THE BOUNDED DERIVATIVE PRINCIPLE

In this section we address the question of how long a solution will continue to vary on the slow time scale after initialization. Kreiss (1978, 1979, 1980) has considered this question for both ordinary and partial differential equations and has formulated an easily stated principle, the bounded derivative principle. We will explain this principle for a system of ordinary differential equations (ODEs) with two time scales, give an example demonstrating the correctness of the principle, and illustrate its application to a system of partial differential equations (PDEs) of interest in NWP.

The occurrence of both slow quasi-geostrophic motions and fast inertia-gravity waves in NWP models indicate the presence of two different time scales in the same system of equations. After an appropriate scaling of the equations of motion, the slow waves have time derivatives which are O(1) times their space derivatives and the fast waves have time derivatives which are O(1/ϵ) times their space derivatives, where ϵ is the nondimensional Rossby number (Sec. 3), 0 < ϵ << 1. Both the slow and fast time scales are associated with purely oscillatory behavior: no exponential growth or decay occurs. This fact is important in the application of the bounded derivative principle.

4.1 The bounded derivative principle for systems of ODEs

To present the principle in its simplest form, we consider a system of ODEs with two time scales,

$$d\underline{y}/dt = A(t)\underline{y} + \underline{f}(t), \qquad (4.1)$$

where $\underline{y}$ is an n-vector and A is an n × n matrix. We assume that A(t) is diagonalizable with purely imaginary eigenvalues $\lambda_1,\ldots,\lambda_n$. We also assume that solutions $\underset{\sim}{z}(t)$ of the homogeneous equation

$$d\underset{\sim}{z}/dt = A(t)\underset{\sim}{z} \qquad (4.2)$$

satisfy an estimate of the form

$$|\underset{\sim}{z}(t)| \leq K|\underset{\sim}{z}(0)| . \qquad (4.3)$$

These assumptions imply that solutions of system (4.1) behave like solutions of a hyperbolic system of PDEs, with motions which are essentially oscillatory. We suppose that the eigenvalues split into two sets

$$M_1 = \{\lambda_1,\ldots,\lambda_m\}$$
$$M_2 = \{\lambda_{m+1},\ldots,\lambda_n\}$$

where $\lambda_j = O(1/\varepsilon)$ if $\lambda_j \in M_1$ and $\lambda_j = O(1)$ if $\lambda_j \in M_2$, with $0 < \varepsilon \ll 1$. Eigenvalues with two different orders of magnitude yield motions with two time scales: the motions associated with M_1 are fast, i.e., their time derivatives are $O(1/\varepsilon)$, and the motions associated with M_1 are slow, i.e. their time derivatives are $O(1)$. The initialization problem is to choose y(0) so that the motions with the fast time scale stay small in amplitude.

The bounded derivative principle arises from a simple observation: If $\underset{\sim}{y}(t)$ varies slowly, then its first few time derivatives satisfy

$$\frac{d^\nu \underset{\sim}{y}}{dt^\nu} = O(1), \text{ for } \nu = 1,\ldots,p-1 \qquad (4.4)$$

and some suitable p > 0. In particular, equation (4.4) must hold at time t = 0. This leads to the principle: choose the initial value $\underset{\sim}{y}(0) = \underset{\sim}{y}_0$ so that

$$\frac{d^\nu \underset{\sim}{y}}{dt^\nu} = O(1) \quad \text{at } t = 0 \quad \text{for } \nu = 0,1,\ldots,p-1. \tag{4.5}$$

The result which Kreiss has proved for ODEs and hyperbolic PDEs is that initial values which satisfy the bounded derivative principle generate solutions which vary only on the slow time scale for t in some finite interval $0 \leq t \leq T$, with $T = O(1)$. His results depend on assumptions on the structure of the systems which we will not discuss in detail. Some of these assumptions involve the existence of transformations of the systems into normal forms. Such transformations correspond roughly to the construction of normal modes in nonlinear normal-mode initialization procedures. One important feature of the principle is that these transformations do not have to be carried out in practice, when applying the principle to a given system.

To illustrate the validity of this principle, consider the scalar equation

$$\frac{dy}{dt} = \frac{i}{\varepsilon}(y + e^{it}), \tag{4.6}$$

$$y(0) = y_0.$$

The solution of this equation is

$$y(t) = y_1(t) + y_2(t), \tag{4.7a}$$

where

$$y_1(t) = -\frac{1}{1-\varepsilon} e^{it} \tag{4.7b}$$

is the slow part of the solution, and

$$y_2(t) = \left(y_0 + \frac{1}{1-\varepsilon}\right) e^{it/\varepsilon} \tag{4.7c}$$

is the fast part of the solution. The derivatives of y are, correspondingly,

$$\frac{d^\nu y}{dt^\nu} = -i^\nu \left(\frac{1}{1-\varepsilon}\right) e^{it} + \left(\frac{i}{\varepsilon}\right)^\nu \left(y_0 + \frac{1}{1-\varepsilon}\right) e^{it/\varepsilon}.$$

For each time derivative $d^\nu y/dt^\nu$, $\nu = 0,1,2,\ldots$, to be $O(1)$, we need

$$y_0 = -\frac{1}{1-\varepsilon} + O(\varepsilon^\nu). \tag{4.8}$$

Hence a particular derivative will be dounded, $d^\nu y/dt^\nu = O(1)$ for all t, <u>if and only if</u> $d^\nu y/dt^\nu = O(1)$ at $t = 0$.

The homogeneous equation corresponding to equation (4.6) has solutions which vary only on the fast time scale $O(1/\varepsilon)$. For a solution of the homogeneous equation to have p bounded derivatives, its initial value must be $O(\varepsilon^p)$. Equation (4.6) has really only a fast time scale, so we expect at most one solution of the inhomogeneous equation, to within $O(\varepsilon^p)$, to have p bounded derivatives. Because of the forcing term, this solution $y_1(t)$ is not zero to within $O(\varepsilon^p)$; its scale of motion is like a slow time scale. The choice of y_0 in equation (4.8) depends in fact on the forcing term in equation (4.6).

For this simple example, in which the explicit solution (4.7) is known, the initialization problem can be handled completely by setting

$$y_0 = -\frac{1}{1-\varepsilon}.$$

Then all derivatives of y are bounded. In general, the initialization problem is not as straightforward as this. We now illustrate how equation (4.6) can be used, together with the bounded derivative principle, to derive equation (4.8) without using the explicit solution (4.7).

By differentiating equation (4.6), we can derive equations for $d^\nu y/dt^\nu$ in terms of y, e.g.,

$$\frac{d^2y}{dt^2} = \frac{i}{\varepsilon}\left(\frac{dy}{dt} + ie^{it}\right)$$

$$= \frac{i}{\varepsilon}\left(\frac{i}{\varepsilon}(y + e^{it}) + ie^{it}\right). \tag{4.9}$$

Evaluating (4.9) at $t = 0$,

$$\left.\frac{d^2y}{dt^2}\right|_{t=0} = -\frac{1}{\varepsilon^2}(y_0 + 1 + \varepsilon),$$

we see that $y''(0) = O(1)$ if and only if $y_0 + 1 + \varepsilon = O(\varepsilon^2)$. Differentiating equation (4.9) successively and using equation (4.6) after each differentiation, we obtain

$$\frac{d^\nu y}{dt^\nu} = \left(\frac{i}{\varepsilon}\right)^\nu y + \left(\sum_{j=1}^{\nu}\left(\frac{i}{\varepsilon}\right)^j i^{\nu-j}\right)e^{it}. \tag{4.10}$$

Evaluating (4.10) at $t = 0$, we see that $(d^\nu y/dt^\nu)(0) = O(1)$ if and only if

$$y_0 = -\sum_{k=0}^{\nu-1} \varepsilon^j + O(\varepsilon^\nu), \tag{4.11}$$

which agrees with equation (4.8), being just the power series of $-1/(1-\varepsilon)$.

In systems where there is only a fast time scale, in general only one solution to within $O(\varepsilon^p)$ has p bounded derivatives. An asymptotic expansion in powers of ε for the initial data of this single slow solution can be obtained from the system using the bounded derivative principle as we have just illustrated.

In systems where two time scales are present in the homogeneous equations, the "slow parts" of the initial data can be specified arbitrarily; then the bounded derivative principle can be used to determine an asymptotic expansion for the "fast parts" of the initial data. For example, consider the system

$$\frac{d\underline{y}}{dt} = \frac{d}{dt}\begin{bmatrix} \underline{y}^I \\ \underline{y}^{II} \end{bmatrix} = \begin{bmatrix} \frac{1}{\varepsilon}A_{11} & \frac{1}{\varepsilon}A_{12} \\ A_{21} & A_{22} \end{bmatrix} \begin{bmatrix} \underline{y}^I \\ \underline{y}^{II} \end{bmatrix} + \begin{bmatrix} \frac{1}{\varepsilon}\underline{f}^I \\ \underline{f}^{II} \end{bmatrix}, \quad (4.12)$$

where $\underline{y}^I$ is the fast part, and $\underline{y}^{II}$ the slow part of the solution. Under certain assumptions, Kreiss (1979) has shown that for every $y^{II}(0)$, there is a $y^I(0)$ so that the solution of system (4.12) with these initial values has p bounded derivatives, and that given $y^{II}(0)$, this $y^I(0)$ is unique to within $O(\varepsilon^p)$.

4.2 The bounded derivative principle in NWP

In the systems of PDEs governing NWP models, it is not clear a priori what the "slow and fast parts" of the initial data are. The bounded derivative principle can be used to determine a sequence of diagnostic equations which hold up to a certain order in ε. Initial data which satisfy these diagnostic equations will generate solutions of the model equations which vary on the slow time scale up to a time which is $O(1)$. The question of how initial data which satisfy these diagnostic equations are constructed from asynoptic measurements does not have a simple answer; each pair of model and nonstandard data sets must be analyzed separately.

We conclude this section by presenting a part of the application of this principle to the shallow-water equations with orography $H = H(x,y)$; this work is due to Browning, Kasahara, and Kreiss (1980). The equations are

$$\frac{du}{dt} + \phi_x - fv = 0, \quad (4.13a)$$

$$\frac{dv}{dt} + \phi_y + fu = 0, \quad (4.13b)$$

$$\frac{\partial \phi}{\partial t} + (u\phi)_x + (v\phi)_y + (\phi_0 - \phi)\delta - (u\phi_x + v\phi_y) = 0, \qquad (4.13c)$$

where $d/dt = \partial/\partial t + u\,\partial/\partial x + v\,\partial/\partial y$, ϕ_0 = mean geopotential, ϕ = deviation in geopotential from the mean, $\delta = u_x + v_y$ = divergence, $\zeta = -u_y + v_x$ = vorticity, and $\Phi = gH$. The nondimensionalized and scaled equations for the case of a midlatitude beta-plane become

$$\frac{du}{dt} + a = 0, \qquad (4.14a)$$

$$\frac{dv}{dt} + b = 0, \qquad (4.14b)$$

$$\frac{\partial \phi}{\partial t} + (u\phi)_x + (v\phi)_y + c = 0, \qquad (4.14c)$$

where

$$a = \varepsilon^{-1}(\phi_x - fv), \qquad (4.15a)$$

$$b = \varepsilon^{-1}(\phi_y + fu), \qquad (4.15b)$$

$$c = \varepsilon^{-2}(\phi_0 - \varepsilon\Phi)\delta - \varepsilon^{-1}(u\phi_x + v\phi_y), \qquad (4.15c)$$

and $f = f_0 + \varepsilon\beta y$, ε = Rossby number = $O(1/10)$. When we say in the sequel that a function is $O(1)$, we mean that the function and its spatial derivatives are $O(1)$.

The first-order time derivatives u_t, v_t, and ϕ_t are $O(1)$ if and only if a, b, and c are $O(1)$. Differentiating (4.15a) with respect to x and (4.15b) with respect to y, this requires

$$a_x + b_y = \varepsilon^{-1}(\nabla^2\phi - f\zeta + \varepsilon\beta u) = O(1), \qquad (4.16a)$$

$$c = \varepsilon^{-2}(\phi_0\delta - \varepsilon\Phi\delta - \varepsilon(u\phi_x + v\phi_y)) = O(1), \qquad (4.16b)$$

i.e.,

$$\nabla^2 \phi - f\zeta + \varepsilon\beta u = O(\varepsilon) \, , \tag{4.17a}$$

$$\delta = \frac{\varepsilon(u\phi_x + v\phi_y)}{\phi_0 - \varepsilon\phi} + O(\varepsilon^2) \, . \tag{4.17b}$$

Dropping terms $O(\varepsilon)$ in (4.17a) and $O(\varepsilon^2)$ in (4.17b), we obtain the diagnostic equations

$$\nabla^2 \phi - f\zeta = 0 \, , \tag{4.18a}$$

$$\delta = \varepsilon(u\phi_x + v\phi_y)/\phi_0 \, . \tag{4.18b}$$

Equation (4.18b) implies that $\delta = O(\varepsilon)$, so the divergent part of the wind is $O(\varepsilon)$. Hence it is permissible to replace u and v by their rotational parts u^0 and v^0 in equation (4.18b), making an error of $O(\varepsilon^2)$. If the rotational part of the wind is measured, which is equivalent to the vorticity ζ being measured, then δ can be determined by (4.18b) and ϕ by (4.18a). Then u and v can be determined from δ and ζ by the Helmholtz theorem.

Notice that equation (4.18a) is the linear balance equation. The numerical experiments of Browning et al. (1980) indicate that requiring just the first time derivatives of u, v, ϕ to be O(1) at time t = 0 does not effectively prevent the growth of inertia-gravity waves. The improved diagnostic relations obtained by requiring that the second time derivatives of u, v, ϕ also be O(1) at time t = 0 do effectively prevent the growth of the first waves. These results corroborate those obtained in the practical implementation of nonlinear normal-mode initialization schemes (Daley, 1980).

We will not present the second-order relations here. In particular, the nonlinear balance equation can be derived from these second-order relations (compare also Sec. 3, and Leith (1980)). The reader is referred to Browning et al. (1980) for details.

5. CONCLUDING REMARKS

We have pointed out that there are two problems in passing from a set of measured data, distributed in space and time, to a solution of the governing equations for an NWP model: one is completeness, the other is, for lack of a better term, accuracy. We call a set of measured synoptic and asynoptic data <u>complete</u> if it determines uniquely the solution to the model equations, and if small errors in the data lead only to small errors in this solution. Criteria were outlined for the theoretical completeness of data sets considering some very simple model equations.

Given a complete data set, we studied the practical construction of the unique model solution by <u>intermittent</u> <u>updating</u>. The convergence of updating procedures was investigated, and its dependence on the updating interval was stressed.

The second problem, that of <u>accuracy</u>, is linked to the presence of two time scales, a slow and a fast one, in the model equations of NWP. For a model system, the <u>shallow-water</u> <u>equations</u>, we showed that the geopotential field and its first two time derivatives, or tendencies, form a complete set of data. The velocity field can be constructed by solving a set of <u>diagnostic</u> equations which use the mass field data. It was shown that the initial state thus constructed generates a slow solution to the prognostic model equations, provided the tendencies of the mass field at initial time are small. Hence this solution of the completeness problem also solves the <u>initialization</u> <u>problem</u>, that of ascertaining that the evolution of the model is on the slow time scale only.

It remained to show that, given small initial tendencies of the solution, the solution will continue to evolve slowly for a time interval comparable to the model's slow time scale. This is in fact the case, as shown by the <u>bounded</u> <u>derivative</u> <u>principle</u>. This principle was first illustrated for simple systems of ODEs with two time scales. It was then applied to the shallow-water equations.

In conclusion, we have addressed certain theoretical questions of 4-D data assimilation and initialization: completeness of data sets, convergence of intermittent updating, the time span of slow evolution of NWP model solutions. The questions and their solution were illustrated for the shallow-water equations.

We hope to have shown the close relationship between data assimilation and initialization. A unified treatment of both aspects of the data problem in NWP appears in the companion paper, Ghil <u>et al</u>. (1980).

Acknowledgements

It is a pleasure to acknowledge useful discussions and correspondence with F. Baer, L. Bengtsson, M. Halem, E. Isaacson,

E. Källén, H.-O. Kreiss, C. E. Leith, J. Oliger and O. Talagrand. Work on this review article was supported by the National Aeronautics and Space Administration under grants NSG-5034 (K. B.) and NSG-5130 (M. G.), monitored by the Goddard Laboratory for Atmospheric Sciences.

References

Baer, F., and J. J. Tribbia, 1977. On complete filtering of gravity modes through nonlinear initialization. Mon. Wea. Rev., 105, 1536-1539.

Bengtsson, L., 1975. 4-Dimensional Assimilation of Meteorological Observations. GARP Publications Series, No. 15, World Meteorological Organization, Geneva, 76 pp.

Browning, G., A. Kasahara, and H.-O. Kreiss, 1980. Initialization of the primitive equations by the bounded derivative method. J. Atmos. Sci., 37, 1424-1436.

Bube, K. P., 1978. The construction of initial data for hyperbolic systems from nonstandard data. Ph.D. Dissertation, Department of Mathematics, Stanford University, Stanford, CA 94305, 114 pp.

_____, 1980. Determining solutions of hyperbolic systems from incomplete data. Comm. Pure Appl. Math., submitted.

_____ and J. Oliger, 1977. Hyperbolic partial differential equations with nonstandard data, in Advances in Computer Methods for Partial Differential Equations, II, R. Vichnevetsky (ed.), IMACS, Rutgers Univesity, New Brunswick, NJ 08903, pp. 256-263.

Charney, J., M. Halem, and R. Jastrow, 1969. Use of incomplete historical data to infer the present state of the atmosphere. J. Atmos. Sci., 26, 1160-1163.

Daley, R., 1980. Normal mode initialization, this volume.

Chi1, M., 1975. Initialization by compatible balancing, Report No. 75-16, Institute for Computer Applications in Science and Engineering, NASA Langley Research Center, Hampton, VA 23665, 38 pp.

_____, 1980. The compatible balancing approach to initialization, and four-dimensional data assimilation. Tellus, 32, 198-206.

_____, and R. Balgovind, 1979. A fast Cauchy-Riemann solver. Math. Comput., 33, 585-635.

_____, B. Shkoller and V. Yangarber, 1977. A balanced diagnostic system compatible with a barotropic prognostic model, Mon. Wea. Rev., 105, 1223-1238.

_____, S. Cohn, J. Tavantzis, K. Bube and E. Isaacson, 1980. Applications of estimation theory to numerical weather prediction, this volume.

Kreiss, H.-O., 1978. Problems with different time scales. In Recent Advances in Numerical Analysis, C. de Boor and G. H. Golub (eds.), Academic Press, New York, pp. 95-106.

_____, 1979. Problems with different time scales for ordinary differential equations. SIAM J. Numer. Anal. 16, 980-998.

_____, 1980. Problems with different time scales for partial differential equations. Comm. Pure Appl. Math. 33, 399-439.

Leith, C. E., 1980. Nonlinear normal mode initialization and quasi-geostrophic theory. J. Atmos. Sci. 37, 958-968.

Machenhauer, B., 1977. On the dynamics of gravity oscillations in a shallow water model, with application to normal mode initialization. Beitr. Phys. Atmos. 50, 253-271.

Smagorinsky, J., K. Miyakoda and R. F. Strickler, 1970. The relative importance of variables in initial conditions for dynamical weather prediction. Tellus 22, 141-157.

Talagrand, O., 1980. A study of the dynamics of four-dimensional data assimilation. Tellus, to appear.

APPLICATIONS OF ESTIMATION THEORY TO NUMERICAL WEATHER PREDICTION

M. Ghil[*], S. Cohn, J. Tavantzis[**], K. Bube, and E. Isaacson

Courant Institute of Mathematical Sciences
New York University, New York, New York 10012

Abstract

Numerical weather prediction (NWP) is an initial-value problem for a system of nonlinear partial differential equations (PDEs) in which the initial values are known only incompletely and inaccurately. Data at initial time can be supplemented, however, by observations of the system distributed over a time interval preceding it. Estimation theory has been successful in approaching such problems for models governed by systems of ordinary differential equations and of linear PDEs. We develop methods of sequential estimation for NWP.

A model exhibiting many features of large-scale atmospheric flow important in NWP is the one governed by the shallow-fluid equations. We study first the estimation problem for a linearized version of these equations. The vector of observations corresponds to the different atmospheric quantities measured and space-time patterns associated with conventional and satellite-borne meteorological observing systems. A discrete Kalman-Bucy (K-B) filter is applied to a finite-difference version of the equations, which simulates the numerical models used in NWP.

---

[*] Also Laboratory for Atmospheric Sciences,
NASA Goddard Space Flight Center, Greenbelt, MD 20771

[**] Department of Mathematics, New Jersey Institute of Technology,
Newark, N.J. 07102

The specific character of the equations' dynamics gives rise to the necessity of modifying the usual K-B filter. The modification consists in eliminating the high-frequency inertia-gravity waves which would otherwise be generated by the insertion of observational data. The modified filtering procedure developed here combines in an optimal way dynamic initialization (i.e., elimination of fast waves) and four-dimensional (space-time) assimilation of observational data, two procedures which traditionally have been carried out separately in NWP. Comparisons between the modified filter and the standard K-B filter have been made.

The matrix of weighting coefficients, or filter, applied to the observational corrections of state variables converges rapidly to an asymptotic, constant matrix. Using realistic values of observational noise and system noise, this convergence has been shown to occur in numerical experiments with the linear system studied; it has also been analyzed theoretically in a simplified, scalar case. The relatively rapid convergence of the filter in our simulations leads us to expect that the filter will be efficiently computable for operational NWP models and real observation patterns.

Our program calls for the study of the asymptotic filter's dependence on observation patterns, noise levels, and the system's dynamics. Furthermore, the covariance matrices of system noise and observational noise will be determined from the data themselves in the process of sequential estimation, rather than be assigned predetermined, heuristic values. Finally, the estimation procedure will be extended to the full, nonlinear shallow-fluid equations.

CONTENTS

1. Introduction
2. A Review of the State-Space Approach to Estimation
    2.1 Statistical considerations in estimation: a simple illustration
    2.2 Estimation for stochastic-dynamic systems
    2.3 General remarks on the Kalman-Bucy (K-B) filter
3. Estimation for the Shallow-Water Equations
    3.1 The equations
    3.2 Discretization
    3.3 The modified K-B filter
    3.4 Observational pattern and choice of parameters
    3.5 Previous work
4. Results
    4.1 Estimation for a perfect, noise-free model
    4.2 Estimation in the presence of system noise
    4.3 Theoretical analysis of the scalar case
5. Concluding Remarks
    References

    Appendix A. List of Major Symbols

1. INTRODUCTION

One of the main reasons we cannot tell what the weather will be tomorrow is that we do not know what the weather is today. In other words, numerical weather prediction (NWP) is an initial-value problem for which initial data are not available in sufficient quantity and with sufficient accuracy.

Numerical forecasts are produced now routinely on a daily basis by a number of weather services in different countries. The models used in NWP are discretized versions of the partial differential equations (PDEs) governing large-scale atmospheric flow. The discretization is performed by finite differencing, finite element or spectral representations. The number of degrees of freedom of the discretized models is typically of the order of 10^5-10^6. The spatial domain of the models is the entire globe or at least an entire hemisphere.

A large number of observations is made by the conventional, ground-based meteorological network, coordinated by the World Weather Watch (WWW). They consist of point values of temperature, humidity, pressure and horizontal velocity. These observations, of the order of 10^5 in number, are produced at the so-called synoptic times, 0000 GMT and 1200 GMT. It is customary therefore to choose a synoptic time as initial time for a numerical forecast. Conventional observations are insufficient in number in order to determine the initial state of the model atmosphere. Furthermore, they are very unevenly distributed in space, being concentrated over the continents of the Northern Hemisphere, and much sparser over the oceans and over the Southern Hemisphere (Fig. 1).

A number of additional observations are made at the so-called subsynoptic times, 0600 GMT and 1800 GMT. A still larger number of observations, exceeding by now that given by the WWW network, is gathered in an essentially time-continuous manner from polar-orbiting satellites and other non-conventional measuring platforms (Fleming et al., 1979 a,b, and references therein). All these observations together form a rather bewildering array by their uneven distribution in space and time, as well as by their different error characteristics.

In order to obtain the best possible estimate of the model state at the initial instant, NWP centers use the data available over a time interval preceding that instant. The most common procedure to use such data, called updating, was suggested by Charney et al. (1969). The model is provided the best available data at some preceding instant, e.g., 24 h or 48 h earlier, and is integrated forward in time. Additional data replace the model values as they become available: the model is updated. When the model integration reaches the initial instant for the next scheduled forecast, its guess of the initial state is blended with the data available at that instant to produce the desired estimate. Thus the model itself is used to assimilate the data

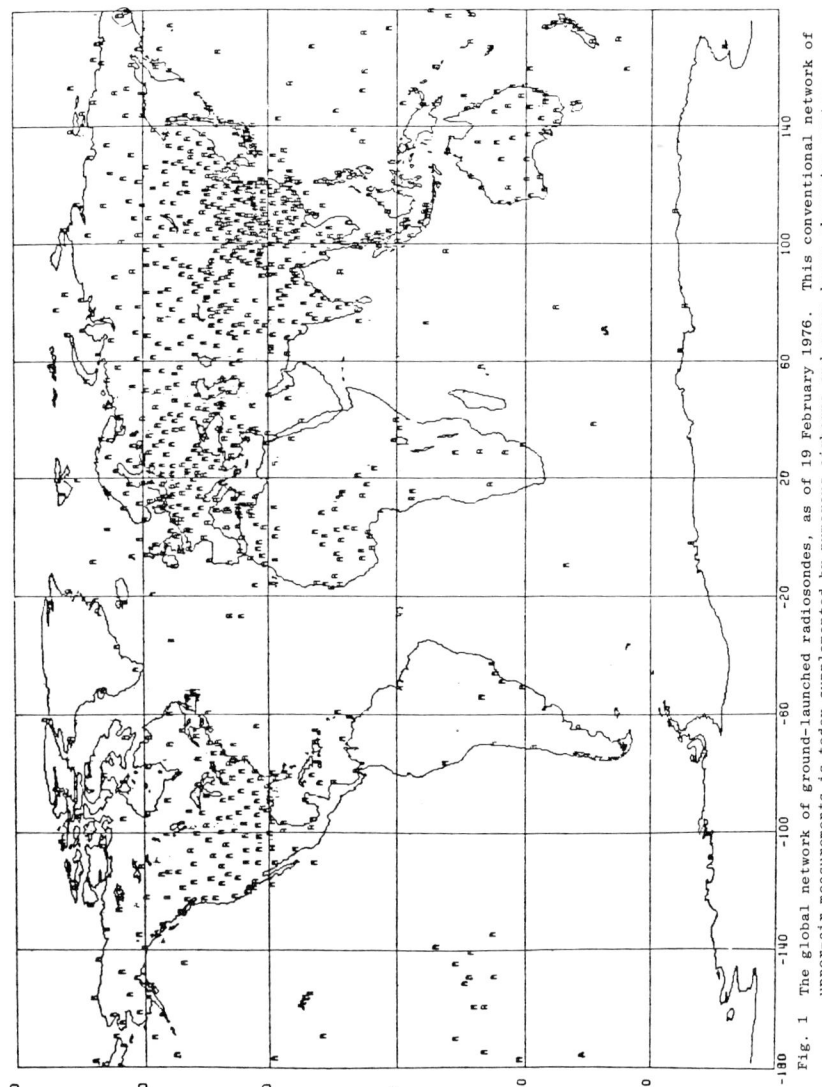

Fig. 1 The global network of ground-launched radiosondes, as of 19 February 1976. This conventional network of upper-air measurements is today supplemented by numerous airborne and space-borne observing systems.

available up to the initial instant. Variations of this procedure, as well as other procedures for the four-dimensional (4-D), space-time assimilation of meteorological observations are reviewed in Bengtsson (1975).

The blending of observations and model forecast values has been made most recently using in an explicit manner the error structure of the data (Phillips, 1976; Rutherford, 1972). This error structure is determined from past data and the resulting linear regression coefficients are computed once and for all, their constant values being used all along the assimilation cycle of the model (McPherson et al., 1979; Schlatter et al., 1976). A modification of this approach, which combines more intimately the dynamics of the model with the time-continuously changing observation patterns, appears in Ghil et al. (1979).

Clearly, a mathematical framework well suited for the 4-D assimilation problem of NWP is the state-space approach of estimation theory. It deals with the estimation of stochastic processes which are generated by randomly perturbed differential or difference equations. This approach was first formulated for processes governed by linear systems of ordinary differential equations (ODEs: Kalman, 1960; Kalman and Bucy, 1961); its results are widely known as the Kalman or the Kalman-Bucy (K-B) filter. We expect that applications of the K-B filter, with suitable modifications, to 4-D data assimilation will provide additional physical insight into this field's outstanding problems and eventually lead to better practical algorithms for solving them.

The purpose of the present article is to apply the K-B filter to the linearized shallow-water equations and to learn as much as possible from this application about the properties of the filter relevant to operational 4-D data assimilation. This linear system already contains important features of the equations used in operational NWP models, and our application should be instructive.

We present a brief review of the K-B filter in Section 2. The dynamical model and observing pattern studied are presented in Section 3, along with the modification of the K-B filter suggested by the system's dynamics. Numerical results and some analytical ones follow in Section 4. A discussion of the results, comparison with operational practice and conclusions follow in Section 5.

2. A REVIEW OF THE STATE-SPACE APPROACH TO ESTIMATION

The intent of this section is to familiarize the reader with the ideas and methods of sequential estimation. Systematic, more or less rigorous expositions of the theory are in existence for the interested reader (Bucy and Joseph, 1968; Curtain and Pritchard, 1978; Davis, 1977; Gelb, 1974; Jazwinski, 1970). Here we shall stay on the purely formal, and hopefully intuitive, level.

2.1 Statistical Considerations in Estimation: A Simple Illustration

Given a quantity x, suppose that two independent measurements of this quantity, x_1 and x_2, are available. For instance x could be the temperature in a room and x_1 and x_2 the readings of two thermometers placed in the room. In the absence of any additional information about x, it is natural to seek an estimate of x, $\hat{x}$ say, as a <u>linear</u> function of x_1 and x_2,

$$\hat{x} = \alpha_1 x_1 + \alpha_2 x_2 . \tag{2.1a}$$

The function itself is called an <u>estimator</u>; the <u>estimate</u> is its value.

We wish to determine α_1 and α_2 so that the estimate $\hat{x}$ will be optimal in some sense. The conditions to achieve such optimality depend on the nature of the measurements.

Let us assume first of all that there are no systematic errors in the measurements x_1 and x_2, i.e. that if we repeat our measurements many times then their average would equal the true value x. In terms of the expectation operator E, this is written as

$$E(x_1 - x) = E(x_2 - x) = 0,$$

and we say the measurements are <u>unbiased</u>. It is natural to require that the estimate $\hat{x}$ also be unbiased:

$$E(\hat{x} - x) = 0;$$

this requirement is equivalent to

$$\alpha_1 + \alpha_2 = 1, \qquad (2.1b)$$

so that (2.1a) becomes

$$\hat{x} = x_1 + \alpha_2(x_2 - x_1). \qquad (2.1c)$$

Next it is assumed that the measurement errors x_1-x and x_2-x are uncorrelated:

$$E(x_1 - x)(x_2 - x) = 0,$$

and that their variances σ_1^2 and σ_2^2,

$$\sigma_1^2 = E(x_1 - x)^2, \quad \sigma_2^2 = E(x_2 - x)^2$$

are known from previous measurements, viz. from instrument calibration. Then the variance of the estimation error, $\hat{\sigma}^2 = E(\hat{x} - x)^2$, is given by

$$\hat{\sigma}^2 = \alpha_1^2 \sigma_1^2 + \alpha_2^2 \sigma_2^2. \qquad (2.2)$$

Suppose for the moment that in addition to satisfying Eq. (2.1b), α_1 and α_2 are nonnegative but otherwise arbitrary; the linear combination in Eq. (2.1a) is then said to be <u>convex</u>. Convexity will imply that $\hat{x}$ always lies between x_1 and x_2, and furthermore that

$$\hat{\sigma}^2 \leq \max\{\sigma_1^2, \sigma_2^2\}. \qquad (2.3a)$$

While property (2.3a) is reassuring, one should be able to do better. We expect to be able to achieve

$$\hat{\sigma}^2 \leq \min \{\sigma_1^2, \sigma_2^2\} ; \qquad (2.3b)$$

otherwise there would be no point in making more than one measurement. The assumed knowledge of σ_1^2 and σ_2^2 will in fact yield (2.3b); it is accomplished by our optimality requirement.

This requirement can now be formulated precisely: $\hat{x}$ should be a <u>minimum variance</u> estimate, which in our case means that $\hat{\sigma}^2$ in Eq. (2.2) should be minimized with respect to α_1 and α_2, subject to (2.1b). The resulting optimal weights are

$$\alpha_1 = \frac{\sigma_2^2}{\sigma_1^2 + \sigma_2^2} = \frac{\hat{\sigma}^2}{\sigma_1^2} \qquad (2.4a)$$

and

$$\alpha_2 = \frac{\sigma_1^2}{\sigma_1^2 + \sigma_2^2} = \frac{\hat{\sigma}^2}{\sigma_2^2} , \qquad (2.4b)$$

where the optimal error variance $\hat{\sigma}$ is given by

$$\hat{\sigma}^{-2} = \sigma_1^{-2} + \sigma_2^{-2} . \qquad (2.4c)$$

Notice that α_1 and α_2 are nonnegative, so that the optimal estimator,

$$\hat{x} = \frac{\sigma_2^2}{\sigma_1^2 + \sigma_2^2} x_1 + \frac{\sigma_1^2}{\sigma_1^2 + \sigma_2^2} x_2 ,$$

is, in fact, convex. Moreover, the optimal weights (2.4a,b) satisfy the intuitive requirement that they should reflect our relative confidence in x_1 and x_2 : if σ_1 is smaller than σ_2, for example, then x_1 is weighted more heavily.

Notice also that formula (2.4c) for the optimal error variance immediately implies property (2.3b). In particular, when $\sigma_1 = \sigma_2 = \sigma$, one obtains $\hat{\sigma} = \sigma/\sqrt{2}$, which generalizes to $\hat{\sigma} = \sigma/\sqrt{N}$ for N independent measurements of equal variance.

2.2 Estimation for Stochastic-Dynamic Systems

The purpose of sequential estimation theory for dynamic systems is to extend the simple ideas outlined above to the case in which the quantity x of interest evolves in time according to a given (system of ordinary or partial) differential or difference equation(s). In this case, x_1 will represent the state of the system as determined from previous observations (measurements), while x_2 represents observations at the current time.

To stress the analogy, let us consider here a system of randomly perturbed <u>difference equations</u> for the state vector $\underset{\sim}{x}$:

$$\underset{\sim}{x}_{k+1} = \Psi \underset{\sim}{x}_k + \underset{\sim}{\xi}_k, \quad k = 0,1,2,\ldots; \qquad (2.5a)$$

$\underset{\sim}{x}$ and $\underset{\sim}{\xi}$ have dimension n, and Ψ is a constant n×n matrix. (See Appendix A for a list of recurring symbols.) In our application, $\underset{\sim}{x}_k$ will stand for the meteorological variables at time k at the grid points of a global atmospheric prediction model; Ψ stands for the finite-difference operator which advances the variables by one time step. The random vector sequence $\{\xi_k: k = 0,1,2,...\}$ is assumed to be a <u>white noise sequence</u> with mean zero and covariance matrix Q,

$$E\xi_k = 0, \quad E\xi_k\xi_\ell^T = Q\delta_{k\ell}. \qquad (2.5b,c)$$

The transpose of a vector or matrix ( ) is indicated by ( )^T and $\delta_{k\ell}$ is the Kronecker delta, $\delta_{k\ell} = 0$ if $k \neq \ell$ and $\delta_{k\ell} = 1$ if $k = \ell$. The white noise ξ represents dynamical and physical processes not described by the model Ψ, especially smaller-scale phenomena not resolved by the grid.

Suppose for the moment that an initial unbiased estimate $\hat{x}_0 = Ex_0$ is available from observations at time zero, and that no further observations are available at later times. In this case the best estimate of x_k, $\hat{x}_k$ say, would be the mean of x_k, $\hat{x}_k = Ex_k$, and is computed according to (2.5), by the recursion

$$\hat{x}_{k+1} = \Psi\hat{x}_k, \quad \hat{x}_0 = Ex_0 ;$$

this estimate can be improved if further observations become available.

We have seen in the introduction that the description of the atmospheric state at a given synoptic time from observations at that time is entirely inadequate and that we are interested in using observations at other times as well. This situation is modeled by assuming a continuous stream of observations

$$z_k = Hx_k + \zeta_k, \quad k = 1,2,3,... ; \qquad (2.6a)$$

ζ_k models observational errors and is also assumed to be a white noise sequence,

$$E\zeta_k = 0, \quad E\zeta_k\zeta_\ell^T = R\delta_{k\ell}, \qquad (2.6b,c)$$

uncorrelated with $\underset{\sim}{\xi}$,

$$E\underset{\sim}{\zeta}_k \underset{\sim}{\xi}_\ell^T = 0 . \tag{2.6d}$$

The zero-mean assumptions (2.5b, 2.6b) are made for convenience only; they are not essential for the theory we describe here.

The observation vectors $\underset{\sim}{z}_k$ have dimension $p \leq n$, and H is a p×n matrix. This formulation describes in particular the meteorological situation in which the observations at any time k are incomplete; measurements are made over some areas (continents) and not over others (oceans); at a given location some variables are measured and not others, e.g., geostationary satellites determine winds but not temperatures. Finally, some functional of the variables may be observed rather than a variable itself, e.g., polar-orbiting satellites measure radiances, which depend on vertical temperature profiles. Thus the entries of H need not be only zero or one, i.e., H is not necessarily a permutation matrix.

In fact, H, as well as Ψ, Q, and R, need not be constant in time; they are only taken constant here for simplicity. In particular, the rank of H, i.e., the effective dimension of $\underset{\sim}{z}_k$, can change from one time to the next: the largest number of observations are available at synoptic times, with fewer observations provided at subsynoptic times, and even fewer in between, at intermediate times.

The <u>linearity</u> assumption that Ψ and H are independent of $\underset{\sim}{x}$ is, however, important for the theory we shall use here. This assumption is essential in guaranteeing the optimality of the filtering algorithm (2.11) below. Extensions to the nonlinear case will be discussed in Sec. 5.

Having described our <u>stochastic-dynamical model</u> (Ψ,H),

Eqs. (2.5, 2.6), we are now in a position to make the connection with (2.1-2.4). Given an estimate $\hat{x}_k(+)$ based on all the observations up to and including the time k, the best <u>prediction</u> at time k+1, $\hat{x}_{k+1}(-)$, is simply

$$\hat{x}_{k+1}(-) = \Psi \hat{x}_k(+) ; \qquad (2.7)$$

$\hat{x}_{k+1}(-)$ will be the analogue of x_1 in Section 2.1. The analogue of x_2 is the actual observation z_{k+1} at time k+1. We wish to combine $\hat{x}_{k+1}(-)$ and z_{k+1} in order to obtain an <u>estimate</u> $\hat{x}_{k+1}(+)$, which we require to be: (a) <u>linear</u>, (b) <u>unbiased</u>, and (c) <u>optimal</u> in some suitable sense.

In Sec. 2.1, we discussed the simple illustrative example of estimating the room temperature x from the readings x_1 and x_2 of two thermometers. In that situation, the requirements (a) and (b) lead to formulae (2.1).

The optimality condition was expressed in (2.4), which yields the minimum $\hat{\sigma}$ among all linear, unbiased estimators.

For our dynamic system (Ψ, H), requirement (a) leads to the formula

$$\hat{x}_{k+1}(+) = L_{k+1}\hat{x}_{k+1}(-) + K_{k+1}z_{k+1}, \qquad (2.8a)$$

which is analogous to (2.1a), while requirement (b) leads to

$$L_{k+1} = I - K_{k+1}H, \qquad (2.8b)$$

which is analogous to (2.1b). Our estimator is therefore of the form

$$\hat{x}_{k+1}(+) = \hat{x}_{k+1}(-) + K_{k+1}(z_{k+1} - H\hat{x}_{k+1}(-)) ; \quad (2.9)$$

the analogy with (2.1c) is obvious. It remains for the gain matrix K_{k+1} to reflect the relative uncertainties in z_{k+1} and $\hat{x}_{k+1}(-)$. This will be achieved by imposing the optimality requirement (c), and will yield formulae analogous to (2.4).

In order to formulate the optimality criterion, we define first the estimation error covariance matrices

$$P_{k+1}(\pm) = E(\hat{x}_{k+1}(\pm) - x_{k+1})(\hat{x}_{k+1}(\pm) - x_{k+1})^T ;$$

$P_{k+1}(-)$ is the counterpart of σ_1^2 and $P_{k+1}(+)$ is that of $\hat{\sigma}^2$. Using Eqs. (2.5) and (2.7), one finds that $P_k(+)$ is advanced by one time step to $P_{k+1}(-)$ according to

$$P_{k+1}(-) = \Psi P_k(+) \Psi^T + Q ; \quad (2.10a)$$

the derivation of (2.10a) depends on the fact that

$$E\xi_k(\hat{x}_k - x_k)^T = 0$$

(cf. (2.5c)). Eqs. (2.6) and (2.9) imply that, in the presence of observations z_{k+1}, $P_{k+1}(+)$ is found from $P_{k+1}(-)$ by the formula

$$P_{k+1}(+) = (I - K_{k+1}H)P_{k+1}(-)(I - K_{k+1}H)^T + K_{k+1}RK_{k+1}^T; \quad (2.10b)$$

in the total absence of observations at time k+1, we have instead $\hat{x}_{k+1}(+) = \hat{x}_{k+1}(-)$ and $P_{k+1}(+) = P_{k+1}(-)$.

The estimation error covariance matrices explicitly contain all relevant information about the error structure of the current estimate.

The statistics of all errors committed up to and including time k are accumulated in $P_k(+)$. Formula (2.10a) shows how this information is advanced to the next time step. For example, this equation determines how the presumably small errors committed over a continent, or other data-rich region, propagate over an ocean, or other data-sparse region. Equation (2.10b) then determines precisely the extent to which the estimate is improved by the new observations.

We have defined the estimation error covariance matrices and considered their changes in time. We are ready now to derive the optimal gain matrix by imposing the optimality requirement (c): it is required that $\hat{x}_{k+1}(+)$ be a <u>minimum variance estimate</u> in the sense that

$$J = E(\hat{x}_{k+1}(+) - x_{k+1})^T S(\hat{x}_{k+1}(+) - x_{k+1})$$

be minimized with respect to each element of K_{k+1}, for all symmetric, positive definite matrices S. In particular, for $S = I$, we see from the definition of $P_{k+1}(+)$ that the trace, or sum of the diagonal elements, of $P_{k+1}(+)$ is to be minimized. The trace of $P_{k+1}(+)$ is the <u>expected mean-square (m-s) estimation error</u>.

We recall that a symmetric matrix S, $S^T = S$, is positive definite if, for any vector $x \neq 0$, the scalar product $x^T S x > 0$. Since every such matrix has a factorization $S = C^T C$, one finds that in general

156

$$J = \text{trace } CP_{k+1}(+) C^T .$$

Using $P_{k+1}(+)$ from Eq. (2.10b) and setting the derivative of J with respect to each element of K_{k+1} equal to zero, one finds that a unique, absolute minimum is attained at

$$K_{k+1} = P_{k+1}(-) H^T (HP_{k+1}(-) H^T + R)^{-1} .$$

This result is valid independently of C, and hence is the same for all possible positive definite matrices S. In other words, K_{k+1} above minimizes simultaneously all reasonable measures, or norms, of the expected estimation error.

The formula above gives the so-called <u>Kalman-Bucy (K-B) optimal gain matrix</u>, or <u>filter</u>. Substituting this into (2.10b) yields the optimal error covariance matrix

$$P_{k+1}(+) = (I - K_{k+1} H) P_{k+1}(-) .$$

Assuming the availability of an unbiased initial estimate

$$\hat{x}_0 = \hat{x}_0(+) = E x_0$$

and an initial estimation error covariance matrix

$$P_0 = P_0(+) = E(\hat{x}_0 - x_0)(\hat{x}_0 - x_0)^T ,$$

the description of the Kalman filtering algorithm is now complete: for $k = 0,1,2,\ldots$, one computes in order

$$\hat{x}_{k+1}(-) = \Psi \hat{x}_k(+) , \qquad (2.11a)$$

$$P_{k+1}(-) = \Psi P_k(+) \Psi^T + Q , \qquad (2.11b)$$

$$K_{k+1} = P_{k+1}(-) H^T (HP_{k+1}(-) H^T + R)^{-1} , \qquad (2.11c)$$

$$P_{k+1}(+) = (I - K_{k+1}H) P_{k+1}(-) , \qquad (2.11d)$$

$$\hat{x}_{k+1}(+) = \hat{x}_{k+1}(-) + K_{k+1}(z_{k+1} - H\hat{x}_{k+1}(-)) . \qquad (2.11e)$$

In the absence of observations at time k+1, Eqs. (2.11c,d,e) are replaced by

$$K_{k+1} = 0 , \qquad (2.11c')$$

$$P_{k+1}(+) = P_{k+1}(-) , \qquad (2.11d')$$

$$\hat{x}_{k+1}(+) = \hat{x}_{k+1}(-) . \qquad (2.11e')$$

Actually, the gain matrix sequence $\{K_k: k = 1,2,3,...\}$ may be precomputed once and for all, i.e., for all realizations of the state and noise processes, x_k, ξ_k, ζ_k. Indeed, (2.11b,c,d) do not depend on the estimates (2.11a,e). This is a result of the assumed linearity of the model (Ψ,H).

To complete the analogy with our earlier illustrative example, notice that Eqs. (2.11c,d) can be rewritten as

$$P_{k+1}^{-1}(+) = P_{k+1}^{-1}(-) + H^T R^{-1} H , \qquad (2.12a)$$

$$K_{k+1} = P_{k+1}(+) H^T R^{-1} \qquad (2.12b)$$

In our analogy, $(P_{k+1}(-), R, P_{k+1}(+), K_{k+1})$ correspond to $(\sigma_1^2, \sigma_2^2, \hat{\sigma}^2, \alpha_2)$; Eqs. (2.12a,b) are analogous to (2.4c) and (2.4b), respectively.

Some intuitively appealing results follow, as in the simple example. For instance, Eq. (2.12a) implies that

$$P_{k+1}(+) \leq P_{k+1}(-) ;$$

the matrix inequality $A \leq B$ means that $C = B - A$ is nonnegative semidefinite, $x^T C x \geq 0$ for all x. Eq. (2.12b) implies that if R is small (large) then the observations z_{k+1}

are weighted more (less) heavily than the predictions $\hat{\underline{x}}_{k+1}(-)$.

2.3 General Remarks on the K-B Filter.

Before proceeding with the description of our dynamical and observational model (Ψ,H), a number of theoretical remarks are in order. Recall first that $\hat{\underline{x}}_{k+1}(+)$ was chosen to be the optimal, minimum variance, unbiased estimate among all estimators of the form (2.8a). It is not clear that in computing $\hat{\underline{x}}_{k+1}(+)$ all past observations $\underline{z}_j$, $j = 1, 2, \ldots, k$, have been fully utilized. It can be shown, however (e.g. Jazwinski, 1970, Sec. 7.3), that our estimate is in fact the optimum unbiased estimate among all estimators which are linear combinations of all the available data

$$\hat{\underline{x}}_{k+1}(+) = A_0 \hat{\underline{x}}_0 + \sum_{j=1}^{k+1} A_j \underline{z}_j . \qquad (2.13)$$

This wider optimality is due to assumptions (2.5c, 2.6c,d) that system errors and observational errors are uncorrelated in time. It can sometimes still be achieved without these assumptions (Jazwinski, 1970, Sec. 7.3, Examples 7.5-7.7). Carrying the estimation error covariance matrices along in the computation makes it possible for the filtering algorithm to be **sequential**, or recursive: each observation is discarded as soon as it is processed. This sequential nature of the estimation makes the algorithm conceptually simple, as well as having great practical advantages. It is one of the major reasons for the broad applicability of Kalman filtering.

Another important feature of the filtering algorithm (2.11) is the fact that only first-order statistics, i.e., means, and second-order statistics, i.e., covariance matrices, of the vector processes of interest are involved. In other words, it suffices to know these statistics for the system noise ξ, observational noise ζ, and initial error $\hat{x}_0 - x_0$. These will provide the estimate $\hat{x}$ as well as its error covariance P at all future times. This property is important since in practice it is usually difficult to obtain even this much information about the random errors one wishes to filter; it is well nigh impossible to obtain more, i.e., to prescribe higher-order statistics. Moreover, the first-order and second-order statistics of the error processes can be determined adaptively, i.e., by the filtering algorithm itself (Chin, 1979, and references therein).

Gaussian processes are in fact completely determined by their first and second-order statistics. Furthermore, the Central Limit Theorem (e.g. Parzen, 1960, Sec. 8.5 and 10.4) states, in its various forms, that the superposition of a large number of random effects is approximately Gaussian, regardless of the distribution of the individual effects. It is reasonable, therefore, to expect our errors, which come from a large number of sources, to be approximately Gaussian. It follows that retention of only first and second-order statistics in the filtering algorithm (2.11) should be rather satisfactory.

Actually, when ξ, ζ, and x_0 are Gaussian, it is known (e.g. Jazwinski, 1970, Sec. 5.2) that the best possible <u>nonlinear</u> estimate of x, i.e., one which might depend nonlinearly on all the observations, is still our linear estimate $\hat{x}$.

The sequential nature of the filter implies in particular that, in the absence of further observations at times $k \geq N$, the best <u>prediction</u> is simply given by (2.11a, 2.11e'):

$$\hat{x}_{k+1} = \Psi \hat{x}_k, \qquad k = N, N+1, N+2, \ldots, \qquad (2.14a)$$

$$\hat{x}_N = \hat{x}_N(+). \qquad (2.14b)$$

The covariance of this predictor is then given by (2.11b, 2.11d'). This corresponds roughly to what is done in operational practice: an initial state $\hat{x}_N$ is determined by 4-D data assimilation from all observations up to and including the synoptic time of interest. Then the forecast model is integrated forward in time from the initial state obtained, without further use of the data. We intend therefore to study only the assimilation and initialization problem, over an assimilation interval $k = 0, 1, 2, \ldots, N$.

The framework of estimation theory can provide also insight into the nature of the system noise (ξ, Q) and ways to its determination. It could lead to improvements in modeling and hence forecasting, by helping to pinpoint deterministic components of ξ. This, however, is not our purpose here. With these remarks, we turn to the description of the dynamic system to which the theory outlined in this section will be applied.

3. ESTIMATION FOR THE SHALLOW-WATER EQUATIONS

3.1 The Equations

The shallow-water equations are a simple system whose solutions exhibit some of the properties of large-scale atmospheric flow. They have certain important characteristics in common with the more complicated, three-dimensional systems currently used in NWP models.

We shall study here a linear, spatially one-dimensional version of the equations, written in cartesian coordinates for a plane tangent to the Earth at latitude θ_0:

$$u_t + Uu_x + \phi_x - fv = 0 , \qquad (3.1a)$$

$$v_t + Uv_x \quad\quad + fu = 0 , \qquad (3.1b)$$

$$\phi_t + U\phi_x + \Phi u_x - fUv = 0 . \qquad (3.1c)$$

The coordinate x points eastward, in the zonal direction, along the circle of latitude $\theta = \theta_0$, while y points northward, in the meridional direction; u and v are velocity components in the x and y directions, $f = 2\Omega \sin \theta_0$ is the Coriolis parameter, with Ω the angular velocity of the Earth. The geopotential $\phi = gh$ measures the deviation of the height $h + H$ of the free surface from its equilibrium value H, with $\Phi = gH$, U is a constant zonal mean flow velocity. All quantities are independent of y.

These equations are derived from the full, nonlinear shallow-water equations on a tangent plane,

$$u_t + uu_x + vu_y + \phi_x - fv = 0 , \qquad (3.2a)$$

$$v_t + uv_x + vv_y + \phi_y + fu = 0 , \qquad (3.2b)$$

$$\phi_t + u\phi_x + v\phi_y + \phi(u_x + v_y) = 0 , \qquad (3.2c)$$

by linearization around a solution ($u = U$, $v = 0$, $\phi = \Phi$) satisfying $fU = -\Phi_y =$ const. We assume in this derivation that the perturbation quantities, i.e., the deviations from equilibrium values of (u,v,ϕ), do not depend on y.

It is advantageous to work with a constant-coefficient system, as long as the basic phenomena of interest are not obscured by this simplification. Hence f, U and Φ in (3.1) are taken to be constants. The variation with latitude of the Coriolis parameter f, however, has an important effect on planetary flows. This so-called β-<u>effect</u>, $\beta = f_y(\theta_0) \neq 0$, is equivalent to the effect of bottom topography, $\Phi_y \neq 0$, in the tangent-plane approximation (Pedlosky, 1979, Sec. 3.17 and Ch. 6). The term $(- fUv)$ in (3.1c) introduces this effect into the solutions of the system considered, without sacrificing the simplicity of constant coefficients.

All the solutions $\underset{\sim}{W} = (u,v,\phi)$ of (3.1) can be expressed as a superposition of plane waves:

$$\underset{\sim}{W}(x,t) = \sum_\ell \underset{\sim}{w}_\ell e^{i\ell(x - c_\ell t)} , \qquad (3.3a)$$

where ℓ is the wave number. For each ℓ, the speed of the individual waves, c_ℓ, is given by the <u>dispersion relation</u>

$$\ell^2(U - c_\ell)\{\Phi - (U - c_\ell)^2\} - f^2 c_\ell = 0 . \tag{3.3b}$$

This is a cubic equation for c_ℓ, which has three <u>real</u> roots, $c_\ell^{(m)}$, $m = 1,2,3$. It turns out that (3.1) has two types of solutions: slow waves, corresponding to

$$c_\ell^{(1)} \cong U - \frac{f^2}{\ell^2 \Phi + f^2} U , \tag{3.4a}$$

and fast waves, with

$$c_\ell^{(2,3)} \cong U \pm \sqrt{\Phi + f^2/\ell^2} + \frac{f^2}{2(\ell^2 \Phi + f^2)} U . \tag{3.4b}$$

In the absence of the β-like term (−fUv) in (3.1c), the last term in (3.4a), as well as in (3.4b), would not be present. The expressions in (3.4) are actually exact to first order in the small nondimensional parameter $U/\sqrt{\Phi}$.

The slow waves are the meteorologically important ones, which correspond to the slow traveling of planetary waves on which synoptic weather systems are superimposed. Their speed is comparable to that of the mean zonal current, $U = O(10 \text{ m/sec})$, and they <u>retrogress</u>: their propagation relative to the mean flow is westward. These slow, retrogressing waves are named after <u>Rossby</u>, and they are an important feature of mid-latitude atmospheric dynamics. Their frequency is always smaller than $f = O(10^{-4} \text{sec}^{-1})$.

The speed of the fast waves is dominated by the second term in (3.4b), being $O(10^2 \text{m/sec})$. They are called <u>inertia-gravity waves</u>, since they are the familiar gravity waves of shallow-water theory, for which $c_\ell = U \pm \sqrt{\Phi}$, modified by the presence of the Coriolis force. Their dispersion and dissipation plays a role in the mechanism of geostrophic adjustment, which maintains the atmosphere in a <u>quasi-geostrophic</u> state, in which the Coriolis force nearly balances the pressure-gradient force. But they carry very little energy at any given time, and appear mostly as higher frequency oscillations superimposed on the meteorologically significant ones, i.e., as <u>meteorological noise</u>.

An important problem in NWP is the <u>filtering</u> of the fast waves in large-scale numerical forecasts, in order to prevent their spurious growth to amplitudes larger than those found in the atmosphere. The need for this filtering is different from that discussed in the previous section: it stems from the two time scales of the deterministic motion itself, rather than from the presence of extraneous, random noise perturbing the deterministic motion. In particular, the fast waves can be eliminated or reduced at the initial time of the forecast by an <u>initialization</u> procedure (Bengtsson, 1975; Leith, 1980; and references therein). Once such an initialization has been performed, the fast waves will not grow excessively over periods of time comparable to the evolution time of the slow waves (Browning <u>et al</u>., 1979; Bube and Ghil, 1980).

Clearly, the two problems of: a) determining a solution to the forecast equations from a continuous stream of noisy data (4-D data assimilation) and b) rendering this solution as free of fast waves as possible (initialization) are related. The connection was stressed and a first step in the direction of their joint solution taken in Ghil (1980) and Leith (1980), among others.

We shall present in the sequel a systematic way of combining the two aspects of filtering within our framework. This will involve a modification of the standard K-B filter outlined in the previous section. Before proceeding with this modification, we shall discretize the equations. This corresponds to what is done in operational practice and will bring the problem to the form (2.5a).

3.2 Discretization

The discretization chosen for (3.1) is in terms of finite differences. Finite-difference models are still the most widely used in NWP. They also facilitate somewhat the assimilation of observations made in irregular patterns. Spectral models, on the other hand, have certain advantages with regard to the initialization aspect of our problem. An analysis similar to the present one should be easy to reproduce for spectral, finite-element or hybrid models.

The grid

$$x^j = j\Delta x, \quad j = 1,2,\ldots,M, \quad t_k = k\Delta t, \quad k = 0,1,2,\ldots \quad (3.5a)$$

is introduced, with $\underset{\sim}{W}_k^j$ approximating $\underset{\sim}{W}(j\Delta x, k\Delta t)$. Then the state vector $\underset{\sim}{x}_k$ of Sec. 2 corresponds simply to $\underset{\sim}{W}_k$, with

$$\underset{\sim}{x}_k = (u_k^1, v_k^1, \phi_k^1, u_k^2, v_k^2, \ldots, \phi_k^M)^T, \quad (3.5b)$$

so that $n = 3M$.

We used the Richtmyer two-step formulation of the Lax-Wendroff scheme (Richtmyer and Morton, 1967, Sec. 12.7 and 13.4). Let A and B denote the matrices of system (3.1),

$$\underset{\sim}{W}_t = A\underset{\sim}{W}_x + B\underset{\sim}{W}.$$

The scheme can then be written as

$$W_{k+1/2}^{j+1/2} = W_k^{j+1/2} + \frac{\Delta t}{2} W_t \Big|_k^{j+1/2}$$

$$= \frac{1}{2}(I + \frac{\Delta t}{2} B)(W_k^j + W_k^{j+1}) + \frac{\Delta t}{2\Delta x} A(W_k^{j+1} - W_k^j), \quad (3.6a)$$

$$W_{k+1}^j = W_k^j + (\Delta t) W_t \Big|_{k+1/2}^j$$

$$= W_k^j + \frac{\Delta t}{\Delta x} A(W_{k+1/2}^{j+1/2} - W_{k+1/2}^{j-1/2}) + \frac{\Delta t}{2} B(W_{k+1/2}^{j-1/2} + W_{k+1/2}^{j+1/2}) \quad (3.6b)$$

Eqs. (3.6) define the dynamics Ψ of our system for the state vector $\underset{\sim}{x}_k$ which is given by (3.5) in terms of $\underset{\sim}{W}_k$,

$$x_{k+1} = \Psi x_k .$$

The discrete system (3.6) has the same type of slow and fast plane wave solutions as the continuous system (3.1). Their dispersive properties are similar. The Lax-Wendroff (L-W) scheme was used because its numerical dissipation is very useful in our simple model in order to simulate the physical dissipation mechanisms active in the atmosphere. Such mechanisms are also present in more complex NWP models, and they are essential for geostrophic adjustment to occur.

The plane waves of the continuous system (3.1) are better approximated by those of the Richtmyer two-step version of the L-W scheme (3.6) than by those of the standard, one-step version. In particular, in the absence of the β-term, the slow, quasi-geostrophic wave of (3.6) satisfies $u_k^j \equiv 0$, as that of (3.1) satisfies $u(x,t) \equiv 0$. This turned out to be a useful check on the departure of solutions w_k^j from geostrophy.

3.3 The Modified K-B Filter

In the absence of any constraints on the dynamics, it is clear from Sec. 2 that the K-B filter corresponds to the solution of an *optimization* problem: obtaining a minimum variance estimator for the system (Ψ, H). We have seen in Sec. 3.1 that, in the application of interest here, it is desirable to select among the solutions of the discrete evolution operator Ψ defined by (3.5, 3.6) a special subset —

solutions with fast components which are vanishing or small. Obtaining a best estimate in the presence of such a constraint corresponds to a <u>constrained</u> optimization problem. We shall study in this subsection an appropriate modification of the standard K-B filter described in Sec. 2.

All solutions of (3.6) can be represented as a superposition of plane waves, similar to (3.3a). For the purposes of this discussion it is actually more convenient to think of $\underset{\sim}{x}_k$ in its physical interpretation $\underset{\sim}{W}_k$, so that we write

$$\underset{\sim}{W}(j\Delta x, k\Delta t) = \sum_{\ell,m} (\lambda_\ell^{(m)})^k \underset{\sim}{w}_\ell^{(m)} \exp\{i\ell(j\Delta x - c_\ell^{(m)} k\Delta t)\}. \quad (3.7a)$$

For each wave number ℓ, there is a slow wave $\underset{\sim}{w}_\ell^{(1)} e^{i\ell j\Delta x}$ with speed $c_\ell^{(1)}$, and two fast waves $\underset{\sim}{w}_\ell^{(2,3)} e^{i\ell j\Delta x}$ with speeds $c_\ell^{(2,3)}$ of opposite signs. The decay factors $\lambda_\ell^{(m)}$, $|\lambda_\ell^{(m)}| \leq 1$, are present due to the dissipation in the difference scheme (3.6).

Denote by R (for Rossby waves), the span of the real parts of all compound n-vectors, $n = 3M$, of the form

$$\begin{Bmatrix} \underset{\sim}{w}_\ell^{(m)} e^{i\ell\Delta x} \\ \underset{\sim}{w}_\ell^{(m)} e^{i\ell 2\Delta x} \\ \vdots \\ \underset{\sim}{w}_\ell^{(m)} e^{i\ell M\Delta x} \end{Bmatrix} \quad (3.7b)$$

for $m = 1$, as ℓ ranges over all possible wave numbers. This is the slow wave space. The fast wave space is denoted by G (for gravity waves), and is defined as the span of the real parts of all n-vectors of the form (3.7b) for $m = 2,3$, as ℓ again ranges over all possible wave numbers.

The spaces R and G are __subspaces__ of Euclidean n-space E^n, and they span E^n, $R \oplus G = E^n$. Furthermore, R and G are invariant under the matrix Ψ, defined by (3.6), i.e., they are invariant subspaces of the system's dynamics. Indeed, the eigenvectors of Ψ are precisely the set of all vectors of the form (3.7b). Hence any vector $\underset{\sim}{x}$ in R will be advanced by the unperturbed system (3.6) to a vector $\Psi\underset{\sim}{x}$ in R after a time step Δt. Similarly, a vector $\underset{\sim}{y}$ in G will evolve to $\Psi\underset{\sim}{y}$ in G. Notice, however, that R and G are __not__ orthogonal to each other.

Given any n-vector $\underset{\sim}{x}$, there is a unique vector $\underset{\sim}{y}$ in R which is closest to $\underset{\sim}{x}$ in the sense that $\|\underset{\sim}{y}-\underset{\sim}{x}\|^2 = (\underset{\sim}{y}-\underset{\sim}{x})^T(\underset{\sim}{y}-\underset{\sim}{x})$ is minimized. This vector $\underset{\sim}{y}$ is called the __orthogonal projection__ of $\underset{\sim}{x}$ onto R, and is denoted by $\underset{\sim}{y} = \Pi\underset{\sim}{x}$. The __orthogonal projection operator__ Π is a symmetric matrix, $\Pi^T = \Pi$, and satisfies $\Pi^2 = \Pi$. The orthogonal projection $\underset{\sim}{y} = \Pi\underset{\sim}{x}$ is found in $O(n \log n)$ arithmetic operations by performing three Fast Fourier Transforms (FFTs) on the vector $\underset{\sim}{x}$ (one for each component $\underset{\sim}{u}$, $\underset{\sim}{v}$, ϕ), then multiplying by a block diagonal matrix comprised of M 3×3 blocks, then performing three inverse FFTs.

We assume in the sequel that

$$\hat{\underset{\sim}{x}}_0 = \Pi \hat{\underset{\sim}{x}}_0 , \qquad (3.8)$$

i.e., that initialization has been performed and $\hat{\underset{\sim}{x}}_0$ lies in R already. These concepts are discussed and similar notation is used in Leith (1980).

Having defined the operator Π, we are now ready to describe the modified filter. Let

$$\underset{\sim}{\eta}_k = \underset{\sim}{z}_k - H \hat{\underset{\sim}{x}}_k(-) \qquad (3.9a)$$

denote the <u>innovation vector</u> at time k. This vector carries the new information contained in the observations at time k; $\hat{\underset{\sim}{x}}_k(-)$ carries all the previously accumulated information, as propagated by the dynamic system.

The filtering step (2.11e) is now written

$$\hat{\underset{\sim}{x}}_{k+1}(+) = \hat{\underset{\sim}{x}}_{k+1}(-) + K_{k+1}\underset{\sim}{\eta}_{k+1} . \qquad (3.9b)$$

What is desired is that, for all k, $\hat{\underset{\sim}{x}}_{k+1}(+)$ lies in R. It is clear by inspection of Eqs. (2.11 a,e,e') that, under assumption (3.8), this will be the case provided that, for all k, $K_{k+1}\underset{\sim}{\eta}_{k+1}$ lies in R. As the observation vector $\underset{\sim}{z}_{k+1}$ is a noisy perturbation of the noisy true state $\underset{\sim}{x}_{k+1}$, cf. Eqs. (2.5a, 2.6a), the <u>correction vector</u> $K_{k+1}\underset{\sim}{\eta}_{k+1}$ does <u>not</u>, in general, lie in R.

What we seek, then, is a modified filter K^*_{k+1}, possibly depending on $\underset{\sim}{\eta}_{k+1}$, which has the property that $K^*_{k+1}\underset{\sim}{\eta}_{k+1}$ <u>does</u> lie in R. This filter is found by minimizing trace $P_{k+1}(+)$, as before, but subject now to the constraint that the correction vector lie in R. A Lagrange multiplier method was used to solve this constrained optimization problem. The result is simply that

$$K^*_{k+1} = \Pi\, K_{k+1} , \qquad (3.10a)$$

independently of $\underset{\sim}{\eta}_{k+1}$, where K_{k+1} is the standard Kalman-Bucy filter. For the modified filter, therefore, we replace Eq. (2.11e) by

$$\hat{x}_{k+1}(+) = \hat{x}_{k+1}(-) + \Pi K_{k+1}\eta_{k+1} \,. \qquad (3.10b)$$

Since this filter is no longer optimal, we must also replace (2.11 d) by (2.10 b), in which K_{k+1} is replaced by ΠK_{k+1}.

In the sequel we shall call the modified algorithm the Π-filter, while the standard algorithm will be called the K-filter. We shall see in Sec. 4 that the Π-filter produces optimal estimates of the _slow_ waves at the expense of estimation errors only slightly larger than the fast-wave contaminated estimates produced by the K-filter.

3.4 Observational Pattern and Choice of Parameters.

In the present article we shall restrict ourselves to the study of a "classical" observational pattern, corresponding to the conventional meteorological upper-air network: all quantities (u,v, ϕ) are observed over "land", and none over the "ocean". This is only meant to serve as an illustration of K-B filtering in a meteorologically familiar situation. Clearly, the power of this approach lies in its ability to handle observations which are arbitrarily distributed in space and time.

Our <u>physical</u> domain is an interval of the x-axis of length 2L. It is meant to correspond to a circle of latitude near 45° N. Hence $f = 10^{-4} \text{sec}^{-1}$ and $L = 14000$ km. The actual distribution of land and ocean at this latitude has been simplified to be 2-periodic, so that in each interval of length L, half is covered by ocean (Pacific or Atlantic), and half by land (North America or Eurasia). It is reasonable to consider, therefore, only 2-periodic solutions of (3.1), and consequently our <u>computational</u> domain is of length L. We consider the left half of the computational L-domain to be covered by land, and the right half to be covered by ocean, so that our observation matrix is

$$H = (I \; 0) \, .$$

The mean flow about which (3.2) is linearized was taken to have $U = 20$ m/s and $\Phi = 3 \times 10^4 \text{m}^2/\text{s}^2$. The value of U is typical for mid-tropospheric flow at this latitude; Φ corresponds to an equivalent depth for a homogeneous atmosphere of $H \cong 3$ km, which gives realistic phase speeds for inertia-gravity waves. The slow waves in the solution of (3.1) which we wish to estimate will travel across the fundamental L-domain of one continent and one ocean in a time of approximately L/U; cf. (3.4a), the actual time, for our choice of L, U, Φ and ℓ, is roughly 12 days.

For a single wave number ℓ, initial data for the continuous system (3.1) which lead only to slow waves are, to first order in the small parameter $U/\sqrt{\Phi}$,

$$\phi(x,0) = \phi_0 \sin \ell x, \qquad (3.11a)$$

$$u(x,0) = \frac{\ell^2 U}{\ell^2 \Phi + f^2} \phi(x,0) = \frac{\ell^2 U \phi_0}{\ell^2 \Phi + f^2} \sin \ell x, \qquad (3.11b)$$

$$v(x,0) = \frac{1}{f} \phi_x(x,0) = \frac{\ell \phi_0}{f} \cos \ell x. \qquad (3.11c)$$

The solution $\underset{\sim}{W}(x,t)$ of (3.1) with initial data $\underset{\sim}{W}(x,0) = \underset{\sim}{W}_0(x)$ given by (3.11) is, to first order in $U/\sqrt{\Phi}$, $\underset{\sim}{W}(x,t) = \underset{\sim}{W}_0(x - c_\ell^{(1)} t)$, with $c_\ell^{(1)}$ given by (3.4a).

We chose initial data corresponding to a single Rossby wave of wave length $L/2$, i.e., $\ell = 4\pi/L$, and amplitude $\phi_0 = 2.5 \times 10^3 m^2/s^2$. The latter is in accordance with a typical ridge-to-trough difference of 500 m in the height of the 500 mb pressure surface (Palmén and Newton, 1969, Sec. 6.6). It follows that $\phi_0/\Phi = 1/12$, which partially justifies the linearization of (3.2). It follows also that the amplitude of $v(x,0)$, $v_{max} = \ell \phi_0/f$, is roughly equal to U, a realistic value. Note, however, that $u_{max} = \ell^2 U \phi_0/(\ell^2 \Phi + f^2)$ is relatively small, $u_{max} \cong 0.053\, v_{max}$, for our choice of parameters.

Initial data for the discrete estimation problem, $\underset{\sim}{\hat{x}}_0(-)$, were computed by evaluating (3.11) at the grid points $x^j = j\Delta x$, while $\underset{\sim}{\hat{x}}_0(+) = \Pi\, \underset{\sim}{\hat{x}}_0(-)$, in accordance with (3.8). The projection is desirable because the slow waves of (3.6) are slightly different from those of (3.1). To obtain $\underset{\sim}{x}_0$, random errors are added to $\hat{x}_0(+)$ (cf. Eqs. (3.12 b,c) below).

Due to the linearity of the problem, the gain matrices K_{k+1}, or ΠK_{k+1}, are independent of both the "true" atmospheric state x and the estimated state $\hat{x}$: Eqs. (2.11a,e) are decoupled from Eqs. (2.11b,c,d). In particular, the gain matrices are independent of the initial data x_0, $\hat{x}_0(+)$. Thus the choice of initial data (3.11) has been made only for orientation purposes, and similar results will obtain for any initial estimate satisfying (3.8).

The discretization used the minimum number of grid points which would resolve our wave, namely 16, for the L-domain of interest. This left a computational problem (2.11) of easily manageable size, and it was deemed sufficient for a preliminary illustration of the method. One numerical experiment was performed with a total of 32 points for the computational L-domain; such a **spatial resolution** of O(400 km) is close to that used in operational NWP models. The results were quite similar to those of the comparable experiment with 16 points.

With a resolution of $\Delta x = L/16$, the stability criterion for the difference scheme (Richtmyer and Morton, 1967, Sec. 12.7) imposes a limit of approximately 47 min on the **time step**, when using (3.4b) for the maximum wave speed. A much more stringent limit on Δt, closer to the value of Δt used in most **primitive-equation** NWP models, in which inertia-gravity waves are present, would be imposed if our spatial resolution Δx were closer to the resolution of such models. We actually took $\Delta t = 30$ min. This results in two time steps per hour, or 24 steps per synoptic period.

Observations are made over the 8 grid points located on "land" at every synoptic time, that is twice per day. The observation error covariance matrix R is taken diagonal, with equal entries at all grid points. The values assigned to the diagonal entries of R are based on data from McPherson et al. (1979, Table 2). The standard deviation of conventional temperature observations used there is 1°C. This can be converted, based on the customary hydrostatic assumption, to a 500 mb level geopotential error of approximately 200 m^2/s^2. This value results, for our choice of ϕ_0, in an error of about 0.1 ϕ_0. A corresponding 10% error in the wind components is roughly 2 m/s; this is slightly larger than the value of 1.5 m/s used by McPherson et al. (1979). We took the standard deviation in observations of ϕ to be 200 m^2/s^2, and that in observations of u and v to be 2 m/s. Relative errors in all observations are thus about 10%.

The system noise covariance matrix Q is taken to be the sum of a geostrophic and an ageostrophic part

$$Q = \Pi D_1^2 \Pi^T + (I - \Pi) D_2^2 (I - \Pi)^T , \qquad (3.12a)$$

with D_1 and D_2 diagonal. The choice of D_1 and D_2 involves further dynamical considerations, and will be discussed after we describe first the experiments with a perfect, noise-free model.

The initial error covariance matrix P_0 also has the form

$$P_0 = \Pi\, D_3^2 \Pi^T + (I - \Pi)\, D_4^2 (I - \Pi)^T , \qquad (3.12b)$$

which results from the assumption that geostrophic and ageostrophic errors are uncorrelated. This assumption is only made for convenience, and because of lack of information on the cross-correlations of the two types of errors; it can be easily removed as further information becomes available. Let D be the diagonal matrix with the elements $(v_{max}, v_{max}, \phi_0)$ repeated on the diagonal; then

$$D_3 = 0.4\, D, \quad D_4 = 0.1\, D . \qquad (3.12c)$$

For simplicity, we have taken the initial error covariances (cf. (3.12 b,c)) to be uniform over the entire L-domain. This choice results in initial errors which are much larger than the observational errors over land, although they are approximately equal to the expected mean forecast errors over the ocean. This uniform distribution of initial errors makes it easier to visualize the initial error reduction by synoptic information over land and the propagation of information from land to ocean, as well as the effect of high errors over the ocean propagating inland. Eventually, we shall be interested in the asymptotic state which corresponds to a stationary, continuous assimilation cycle; hence, the choice of initial errors will ultimately be immaterial.

3.5 Previous Work

The realities of 4-D data assimilation have suggested to practitioners, as well as theoreticians, ideas related to those presented here. A number of authors have preceded us on this ground, and we shall mention their work at least briefly.

Jones (1965a,b) seems to have been the first to bring the formalism of K-B filtering to the attention of the meteorological community. Jones (1965a) is an excellent compendium of the formulae for the discrete-time filter, including results on the asymptotic filter. Jones (1965b) is an attempt at nonlinear filtering for a single, scalar quantity; an improvement over direct insertion obtains when statistical ideas are included.

Petersen (1968, 1970, 1973a,b,c, 1976) offers the most comprehensive treatment of estimation ideas in the meteorological literature. He applied these ideas to a linear form of the quasi-geostrophic barotropic potential vorticity equation (Petersen, 1976), in which fast waves do not appear. The estimation is carried out in terms of spectral transforms and the dynamics incorporated in the form of Green's functions. This approach to estimation does not seem to extend easily to nonlinear dynamics. Its implementation in linear cases is also hampered by not exploiting the sequential aspect of K-B filtering, nor the use of asymptotic filter values.

A variational approach to updating, which bears certain similarities to sequential estimation, appears in Tadjbakhsh (1969) and Phillips (1971). It was also implemented for real satellite sounding data by Ghil and Mosebach (1978). This approach, however, does not include explicitly the statistical optimality considerations of K-B filtering.

Miyakoda and Talagrand (1971) discussed the possibility of blending forecasts from previous synoptic times with current observations, by averaging. They analyzed this possiblity for the linear, one-dimensional vorticity equation and carried out numerical experiments for the same equation in its nonlinear, two-dimensional form. They did not use a sequential filter or statistically determined weights. Still their work showed the importance of using past observations, as carried forward by the dynamical model itself, in obtaining a better estimate for the current state of the atmosphere.

Phillips (1976) developed an estimation procedure for a one-dimensional, linear, two-level quasi-geostrophic model. The model uses fairly realistic flow quantities, observational patterns and error variances. This procedure, like those of Petersen (1976) and of Miyakoda and Talagrand

(1971) is not sequential; it requires, in particular, the specification of the actual solutions' statistics, rather than solely observational and forecast error covariances. Since the model only admits slow waves, no modifications were necessary to eliminate the fast waves present in operational, primitive-equation models.

Phillips' work stressed the importance of statistical concepts in 4-D data assimilation and had a considerable influence in their operational implementation. We hope in turn that our results will lead to a better theoretical understanding of the interaction of statistics and dynamics in meteorological data assimilation and also help practitioners in optimizing further its operational applications.

4. RESULTS

We wish to stress here again that these results are preliminary. They present an illustration of our estimation approach for a very simple model with some nontrivial features of operational NWP models. The main feature of interest is the presence in the model of travelling large-scale waves with different speeds. Optimal estimates of the slow waves are obtained, while eliminating the fast waves.

4.1 Estimation for a Perfect, Noise-Free Model

We study first the way in which the K-filter, using partial observations over land, reduces the initial error in a system without noise, $Q = 0$. Fig. 2 shows the components of the expected root-mean-square (rms) estimation error (trace P_k)$^{1/2}$, over a 10 day numerical experiment, or run. Fig. 2a shows the expected rms error over "land", Fig. 2b over "the ocean", and Fig. 2c over the entire L-domain. The individual curves correspond to the errors in $u(U)$, $v(V)$, $\phi(P)$, and the total error (T).

Obviously, sharp error reduction occurs at the observation times over land (Fig. 2a). The more interesting fact is that noticeable error reduction occurs at synoptic times also over the ocean (Fig. 2b). The latter reduction is due to the corrections applied to ocean grid points by the filter. In this, the K-filter acts like current objective analysis schemes, in particular those using linear regression: the new information

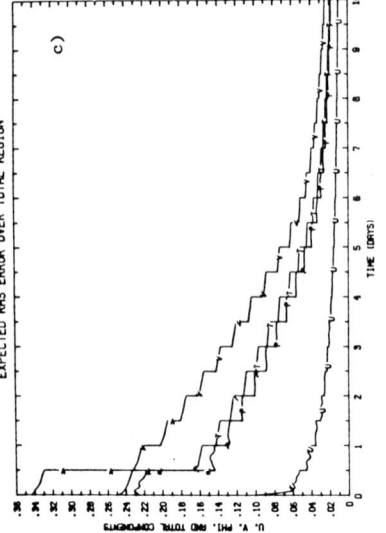
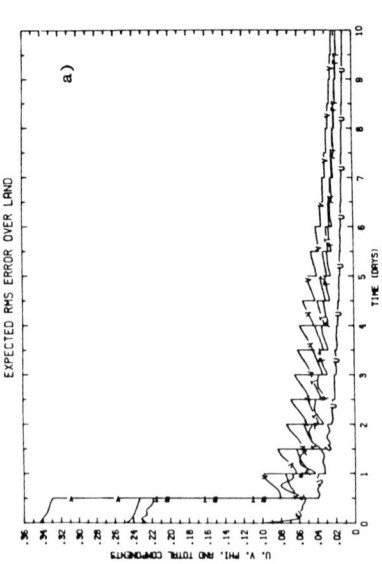
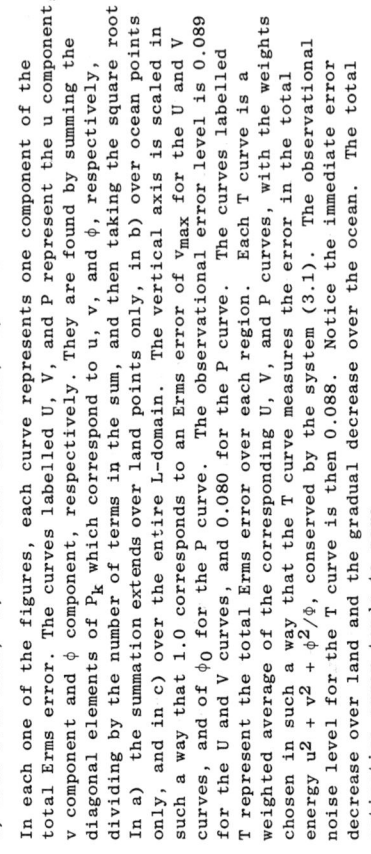

Fig. 2

The components of the total expected rms error (Erms), (trace: $P_k)^{1/2}$, in the estimation of solutions to the stochastic-dynamic system (ψ, H), with ψ given by (3.6) and $H = (I\ 0)$. System noise is absent, $Q = 0$. The filter used is the standard K-B filter (2.11) for the model.

a) Erms over land;   b) Erms over the ocean;   c) Erms over the entire L-domain

In each one of the figures, each curve represents one component of the total Erms error. The curves labelled U, V, and P represent the u component, v component and ϕ component, respectively. They are found by summing the diagonal elements of P_k which correspond to u, v, and ϕ, respectively, dividing by the number of terms in the sum, and then taking the square root. In a), the summation extends over land points only, in b) over ocean points only, and in c) over the entire L-domain. The vertical axis is scaled in such a way that 1.0 corresponds to an Erms error of v_{max} for the U and V curves, and of ϕ_0 for the P curve. The observational error level is 0.089 for the U and V curves, and 0.080 for the P curve. The curves labelled T represent the total Erms error over each region. Each T curve is a weighted average of the corresponding U, V, and P curves, with the weights chosen in such a way that the T curve measures the error in the total energy $u^2 + v^2 + \phi^2/\phi$, conserved by the system (3.1). The observational noise level for the T curve is then 0.088. Notice the immediate error decrease over land and the gradual decrease over the ocean. The total estimation error tends to zero.

from observations over land is spread out over adjacent ocean areas at the observation time itself. The difference between our approach and conventional schemes will become more apparent in the next subsection, inasmuch as the K-filter is capable of discerning between data-sparse ocean areas upstream and downstream from data-rich land.

Even in the present case of a noise-free model, $Q = 0$, the effects of <u>advection of information</u> are noticeable. In between observations, the error over land grows (Fig. 2a). This is due to the advection of error from over the ocean. The total error (Fig. 2c) between observations decreases. This decrease is due to the dissipation in the model, which is conservative when $Q = 0$, except for numerical dissipation. The error over the ocean (Fig. 2b), however, decreases considerably more than the total, due to the advection of information from land.

The expected rms error over land falls below the observational noise level at every observation; over the ocean, this happens after approximately 4-5 days. In our noise-free model, the total expected rms error eventually decays to zero: no information is lost and the observational errors can be eliminated entirely by repeating the observations as long as necessary for the expected estimation error to become negligible. This is a result of the fact that our system is <u>completely observable</u> (Bucy and Joseph, 1968, Ch. 3 and Ch. 5).

Components of the <u>actual</u> rms error in the estimation of an individual realization of our system, with given initial data $\underset{\sim}{x}_0$ and observation error $\{\underset{\sim}{\zeta}_k : k = 1,2,...\}$, appear in Fig. 3a. The corresponding components of <u>relative</u> m-s error, $\|\underset{\sim}{x}_k - \hat{\underset{\sim}{x}}_k\|^2 / \text{trace } P_k$, where $\|\underset{\sim}{y}\|$ is the length of the vector $\underset{\sim}{y}$ are given in Fig. 3b. For a different random choice of $\underset{\sim}{x}_0$ and sequence $\underset{\sim}{\zeta}_k$, the same plots appear in Figs. 3c, 3d.

It is easy to see that the rms error in an individual realization decreases to the observational noise level in about the same time as its expected value (compare Fig. 3a and 3c with Fig. 2c). After that time, however, it continues to fluctuate near zero, rather than decaying monotonically to zero. Our experimental results (Figs. 3b,d) show that the spread of individual estimation errors about their expected value is not too large. It would be interesting, in general, to know <u>a priori</u> how large this spread is expected to be, and we intend to study this question further.

We show in Fig. 4 the actual time histories at a number of grid points for the estimated solution corresponding to the realization in Figs. 3a,b; $\hat{u}$ is shown in Fig. 4a, $\hat{v}$ in Fig. 4b and $\hat{\phi}$ in Fig. 4c. We chose to show a point on the West Coast of the continent, labeled SF (for San Francisco), one on the East Coast, labeled NY (for New York), and one in the middle of the ocean, labeled HA (for Hawaii). Note that "Tokyo" = "New York" by periodicity.

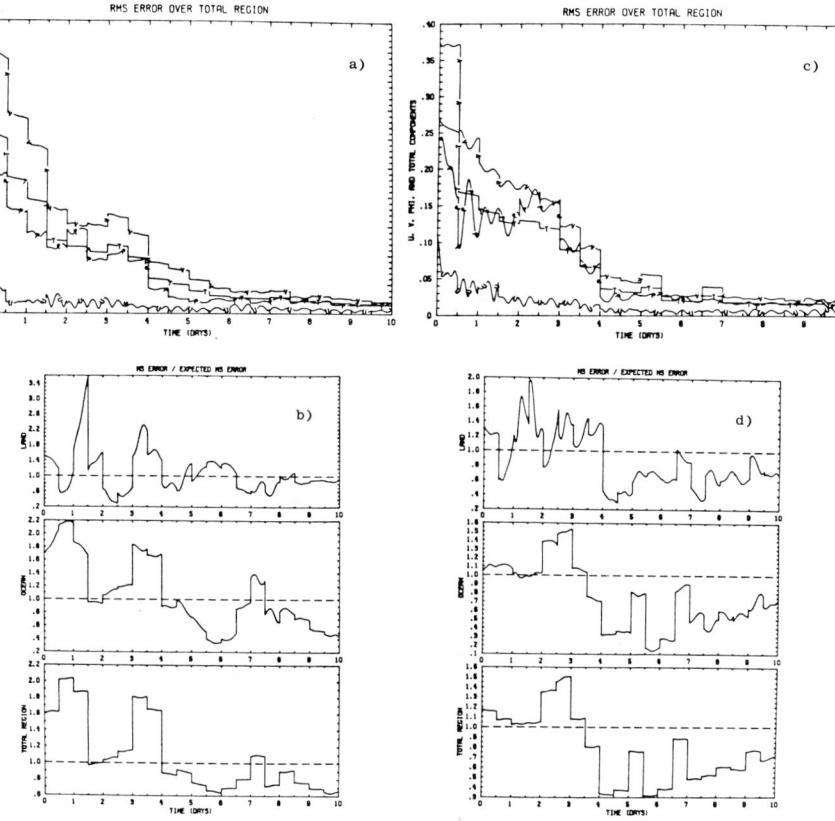

Fig. 3 Components of the actual root-mean-square (rms) estimation error and the relative mean-square (m-s) estimation error, for two realizations of the experiment whose Erms errors appear in Fig. 2. The two realizations correspond to two different random choices of initial condition x_0 and observational noise ζ_k, sampled from the same respective probability distributions, with covariance matrices P_o and R, respectively.

- a) The rms errors for a realization of the run whose Erms errors are given in Fig. 2. The curves have the same labels, and are plotted on the same scale, as in Fig. 2.
- b) Relative m-s errors for the same realization. The three panels give the ratio of total m-s error over land, ocean, and the entire region, respectively, to the corresponding expected m-s error.
- c) Same as Fig. 3a, for a different realization.
- d) Same as Fig. 3b, for the run in Fig. 3c. Notice the difference between Fig. 3a and Fig. 3c: the actual rms estimation error in two realizations of the filtering process can be quite different. Figs. 3b and 3d show the variation of the relative m-s error: equality of absolute m-s error and expected m-s error corresponds to a value of 1 in these figures.

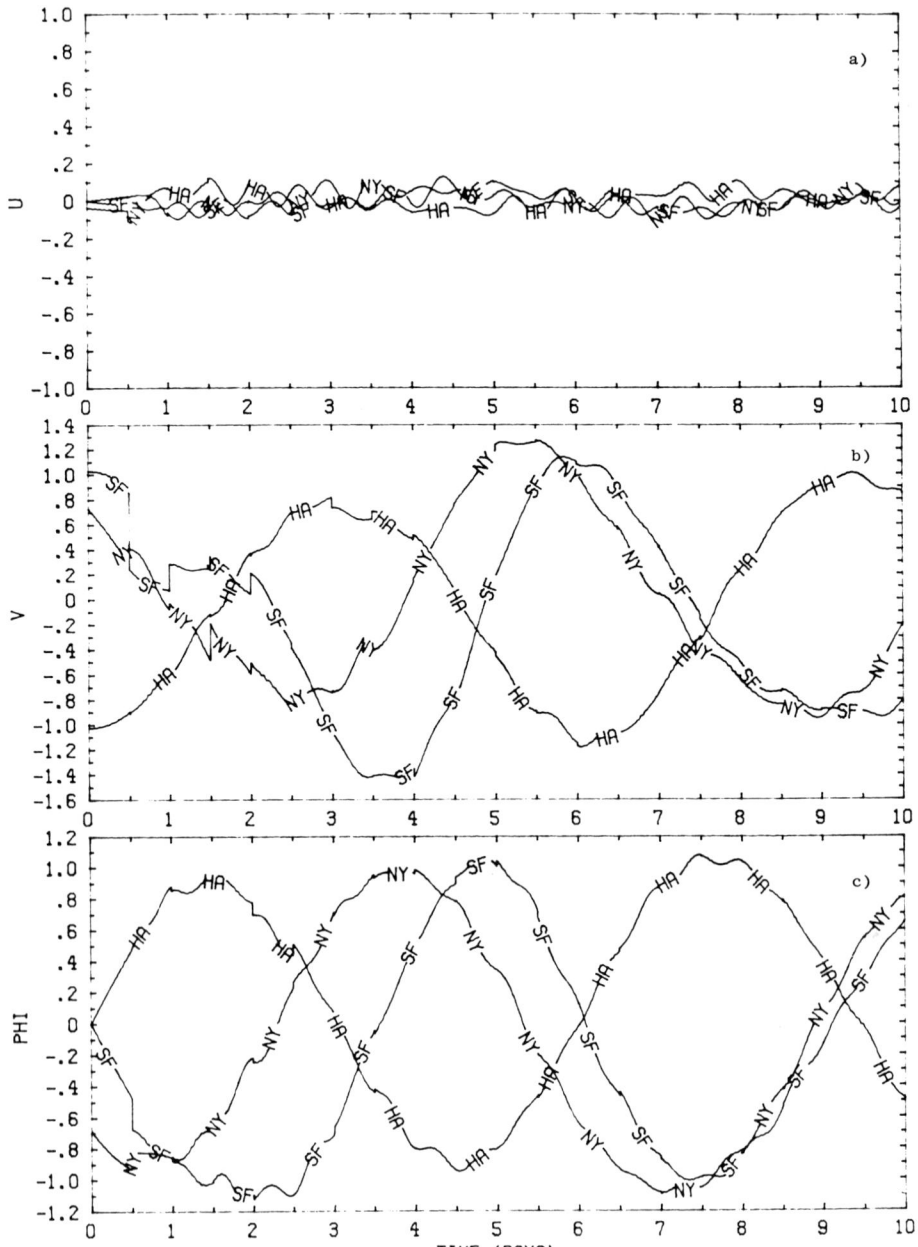

Fig. 4 Time history of the estimated solution without system noise, at three locations, labeled HA (for Hawaii, a mid-ocean location), SF (for San Francisco, a West Coast location), and NY (for New York, an East Coast location).

a) u-component of velocity; b) v-component of velocity,

c) ϕ, the geopotential

Notice the slow waves with a period of approximately 6 days, upon which are superimposed smaller, fast waves with a period of approximately one-half day.

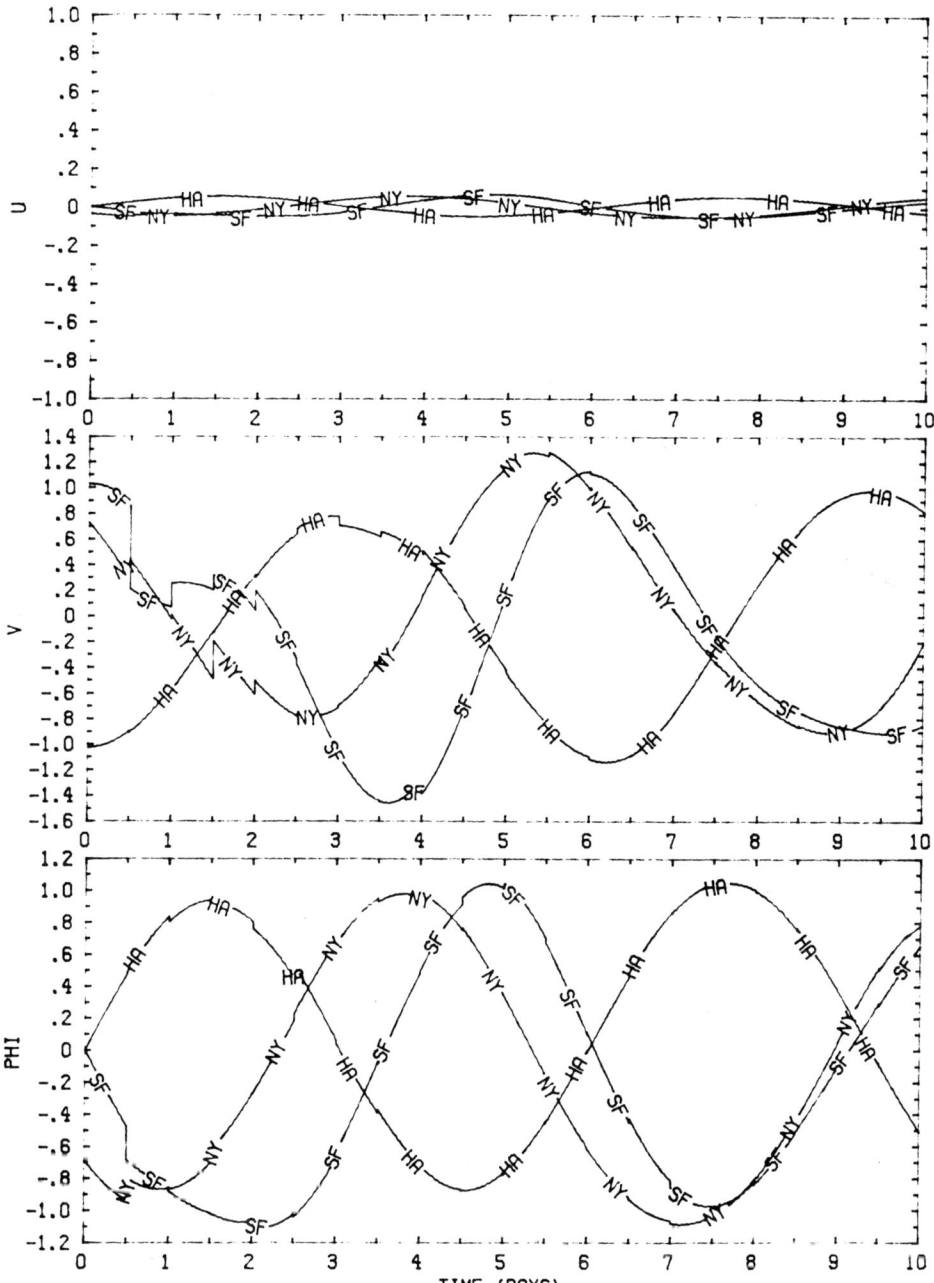

Fig. 5 Same as Fig. 4, only for a run using our modified K-B filter, the Π-filter. The fast waves have been completely eliminated and the estimated solution is slowly varying.

We see that for each curve, small, fast oscillations are superimposed on a smooth, slowly varying wave pattern. The fast oscillations are caused by the ageostrophic component of the initial error, cf. (3.12 b,c), and of the observations. They are especially apparent in the u-component of the estimated solutions. These oscillations are partially damped between synoptic times by the dissipativity of the model.

A run identical in every other respect to that in Fig. 4 was made using the Π-filter instead of the K-filter. The time histories of the corresponding estimates at the same points are shown in Figs. 5a,b,c. They are perfectly smooth, except for the jumps due to observations, which are rather large in the SF and NY curves and very small in the HA curve. In other runs without the β-term (not shown) one had in particular $\hat{u} \equiv 0$, to within machine accuracy. Notice the periodicity of approximately 6 days, due to the passage of the 2-wave we are estimating.

The expected error reduction for the Π-filter (not shown) was only slightly smaller than that for the K-filter. Thus a slowly varying estimated solution was obtained without sacrificing the optimality of the estimate.

We have thus studied error reduction, information propagation and filtering of fast waves in the perfect model. We shall turn our attention presently to the more realistic model with system noise in it.

4.2 Estimation in the Presence of System Noise

In the previous subsection we saw that, for a perfect model, the estimation error covariance matrix $P_k(+)$ tends to zero, and hence so does K_k (Eq. (2.12 b)). We shall study now the case in which, due to the simultaneous presence of system noise Q and observation noise R, the gain matrix K will tend to a nonzero, asymptotic constant value.

4.2.1 Modeling of system noise.

We shall discuss at the beginning our formulation of Q, cf. (3.12a), in particular the choice of the diagonal matrices D_1 and D_2, left open in Sec. 3.4. This choice has to reflect the error growth properties of NWP models. The overwhelming dynamical consideration in this context is the inherent <u>unpredictability</u> of the atmosphere.

Numerous studies (cf. Lorenz, 1969, and references therein) have shown that realistic models of the atmosphere, subject to a small random perturbation, will evolve in a finite amount of time to a state which is <u>statistically independent</u> of the corresponding unperturbed state. This amount of time depends upon the scales of motion of interest, and for synoptic scale motions is about two weeks. Unpredictability is a decidedly <u>nonlinear</u> effect: perturbations at any given scale of motion are nonlinearly fed into <u>all</u> the scales and eventually grow enough to completely contaminate the state.

At that time, forecast error growth levels off (Phillips, 1976). Energy conservation implies in fact that such leveling off must occur in a nonlinear realistic model of the atmosphere.

In our <u>linear</u> model, the estimated state $\hat{\underline{x}}$ is governed, in the absence of observations, by

$$\hat{\underline{x}}_{k+1} = \Psi \hat{\underline{x}}_k \quad , \quad k = 0,1,2,\ldots, \tag{4.1a}$$

while $\underline{x}_k$ is the atmospheric state, governed by

$$\underline{x}_{k+1} = \Psi \underline{x}_k + \underline{\xi}_k \quad , \quad k = 0,1,2,\ldots \tag{4.1b}$$

In this linear model, $\hat{\underline{x}}_k$ and $\underline{x}_k$ will never become actually uncorrelated when they start from the same initial state, $\hat{\underline{x}}_0 = \underline{x}_0$, and hence have the same mean. However, the variance of their difference will grow with time due to the system noise $\underline{\xi}_k$.

We would like to choose Q, i.e., D_1 and D_2 in (3.12a), so as to have

$$\text{trace } P_N = E(\hat{\underline{x}}_N - \underline{x}_N)^T(\hat{\underline{x}}_N - \underline{x}_N) = 2\hat{\underline{x}}_N^T \hat{\underline{x}}_N \tag{4.2}$$

at time N, which would correspond to $\hat{\underline{x}}_N$ and $\underline{x}_N$ being uncorrelated. We prescribed N, in rough agreement with predictability estimates, to be N = 10 days. If the model (4.1a) were conservative, (4.2) would be equivalent to

$$\text{trace } P_N = 2\hat{\underline{x}}_0^T \hat{\underline{x}}_0 \ .$$

In fact, we set $D_1 = \gamma D$, $D_2 = 0.25\gamma D$, similarly to Eq. (3.12c), with γ chosen so as to satisfy

$$\text{trace } P_N = 2\alpha \hat{x}_0^T \hat{x}_0 , \qquad (4.3)$$

and $\alpha = 0.3$. This was easily achieved by trial and error, since P_k evolves simply according to (2.11b, 2.11d'). We found $\gamma = 0.028$ when $\Delta t = 30$ min.

The reason $\alpha = 0.3$ was chosen instead of the $\alpha = 1$ suggested by predictability theory, cf. Eq. (4.2), is that trace P_k in the linear, observation-free model (4.1) continues to grow after time N. It does actually level off also, as a result of dissipation. Its leveling-off time, however, is not related to the predictability limit N and is in fact much larger. Our attempt to account for loss of predictability in our experiments results, therefore, in a choice of Q which, even with $\alpha = 0.3$, is considerably larger than the Q which would appear in a nonlinear model.

We expect actually that the estimation-theoretical framework, applied to experiments with a nonlinear model, will lead to new insights into the nature of atmospheric predictability. One possible approach is the adaptive determination of Q, in which Q is actually determined from the observations themselves (Chin, 1979; Ohap and Stubberud, 1976).

4.2.2 Numerical results. Having prescribed Q, we turn now to the actual experiments with a noisy system. Fig. 6 shows the expected rms error for a run with the K-filter. As in Fig. 2, (a) shows "land", (b) the "ocean", and (c) the entire region.

At the first synoptic time, the total expected rms error (T) over land drops below the observational error level, which in our nondimensional units is 0.088. It grows much more sharply between synoptic times than in Fig. 2a, due to the presence of system noise, which is added to the error advected from the ocean. However, the estimation error just <u>after</u> each synoptic time is smaller than just after the previous synoptic time; the same is true, moreover, of the error just <u>before</u> successive synoptic times.

The <u>monotone decrease</u> of the components of $(\text{trace } P_k(-))^{1/2}$, as well as those of $(\text{trace } P_k(+))^{1/2}$, from one synoptic period to the next is even more striking in Figs. 6b,c. We notice, however, that in contradistinction to Figs. 2b,c, neither the total error (T) over the ocean, nor that over the entire region ever drop below the observational error level. The expected rms errors now increase between synoptic times instead of decreasing: the effect of the system noise $\underset{\sim}{\xi}$ is stronger than the effect of dissipation in Ψ.

What does happen is that the expected rms errors very quickly settle into an asymptotically periodic pattern with the synoptic interval of 12 h as the period. The convergence occurs within 1-2 days over land, and within 4-5 days over the ocean. In particular, the values of trace $P_k(\underline{\pm})$ at synoptic times tend to a constant. This leads us to suspect that in fact the matrices $P_k(\underline{\pm})$ themselves, and hence the filter K_k, tend to a constant.

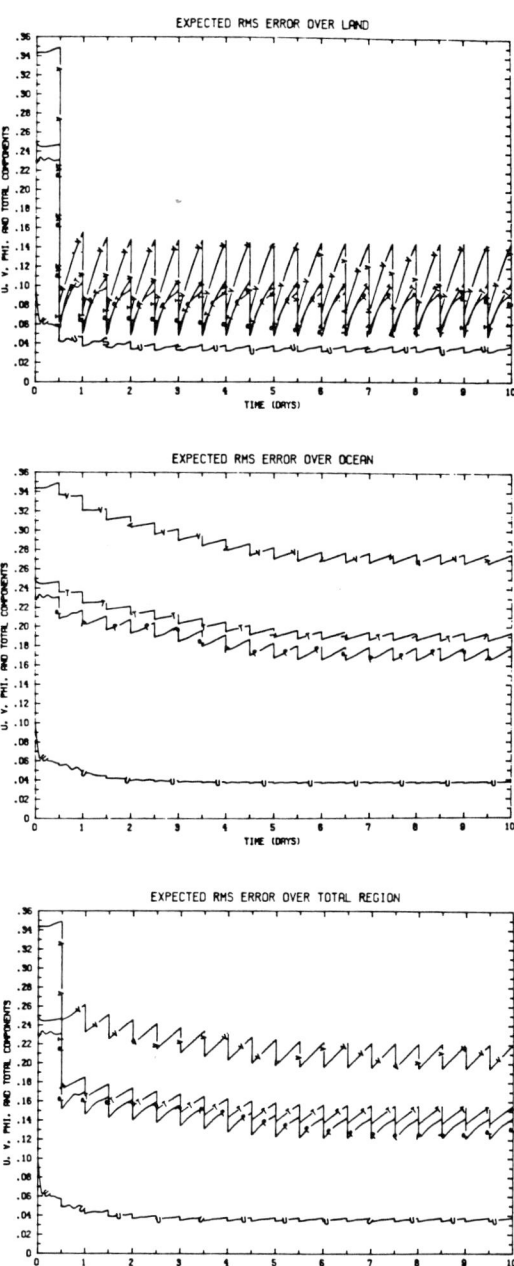

Fig. 0 This figure and the following ones show the properties of the estimated algorithm (2.11) in the presence of system noise, $Q \neq 0$. This figure gives the Erms estimation error, and is homologous to Fig. 2. Notice the sharper increase of error over land between synoptic times, and the convergence of each curve to a periodic, nonzero function.

4.2.3 Asymptotic filter: properties. Examining the behavior of all entries of the gain matrix K as a function of time confirmed our conjecture. Moreover, the asymptotic matrix K_∞, to which K_k tends, is banded. Recall that P is an n × n matrix, with n = 3M = 48, H is a p × n matrix, with p = n/2 = 24, and R is p × p:

$$P = \begin{pmatrix} P_\ell & | & P_{\ell-o} \\ --- & | & --- \\ P_{o-\ell} & | & P_o \end{pmatrix} \quad , \quad H = \begin{pmatrix} I & | & 0 \end{pmatrix} . \quad (4.4a)$$

Here P_ℓ is the submatrix of the auto-covariances for estimation errors over land, $P_\ell^T = P_\ell$, P_o the autocovariance of errors over the ocean, $P_o^T = P_o$, while $P_{\ell-o}$ is the cross-covariance of errors over land and over the ocean, with $P_{o-\ell} = P_{\ell-o}^T$. By (2.12b) and (4.4a),

$$K = \begin{pmatrix} P_\ell R^{-1} \\ P_{o-\ell} R^{-1} \end{pmatrix} , \quad (4.4b)$$

which is n × p.

All entries away from the diagonal of the upper block in K_∞ become rapidly smaller with distance from the diagonal. Periodicity in x leads to the appearance of a few larger elements in the corners of both blocks, the upper and the lower.

To visualize better the behavior of K_k in time and to study the structure of K_∞, we considered contour plots of the elements of K (not shown), and selected cross-sections of these plots. The cross-sections correspond to the influence functions of observations at the selected location. In other words, they show the weight given to an observation at such a location when updating a point situated a certain distance away from it.

The chosen locations, or "upper-air stations", were SF, SL (for Saint Louis) and NY. There is no influence function for mid-ocean points, like HA, since no observations are made there. Cross-sections were plotted at every synoptic time, i.e., every 24 time steps. It was clear that convergence occurred within 4-5 days, as it did for trace P_k over the entire region.

Fig. 7 shows the influence functions for the selected locations at the end of day 10. Fig. 7a marked (u-u) gives the influence of a u observation at the selected stations on u updates at any grid point in the L-domain. Fig. 7b, marked (u-v), gives the weight of a v observation at a station in a u update at every grid point, and so on; Fig. 7i gives the influence of v on ϕ.

All the weighting coefficients involving u are rather small (Figs. 7a,b,c,e,h). This is due to our choice of the

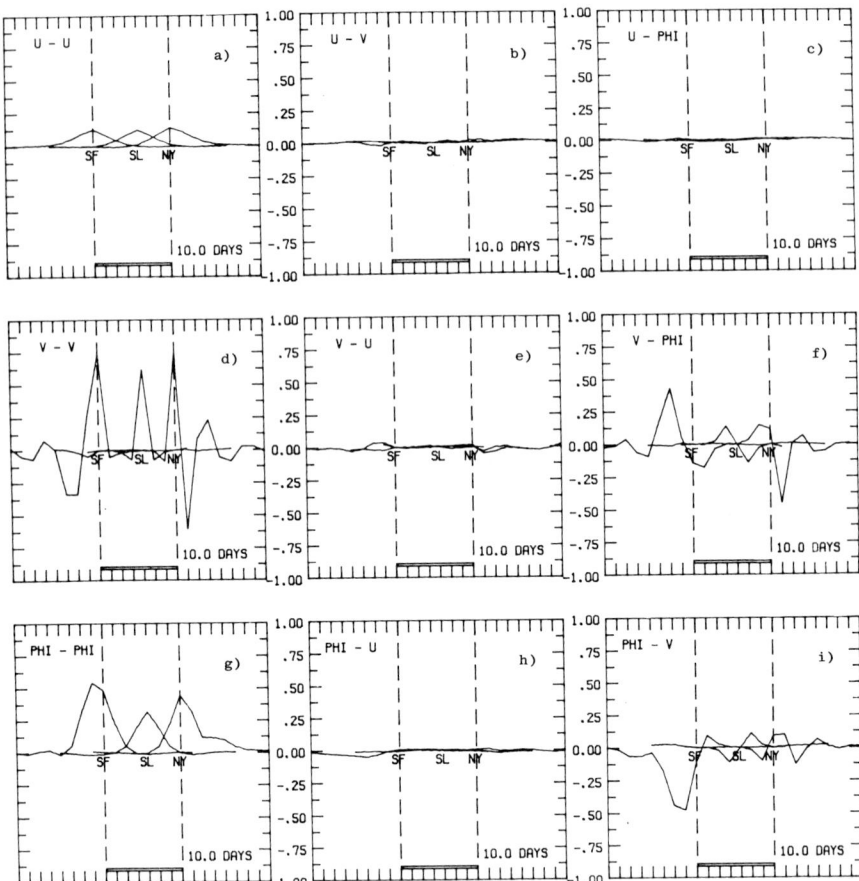

Fig. 7 Influence functions of selected stations at which observations are made. The stations selected are SL (for Saint-Louis, an interior continental location), SF and NY (see Fig. 4). These functions are simply cross-sections of the K-B algorithm's gain matrix K at day 10. They correspond to the weight given an observation at station SF, say, when making a correction to the forecast field at any one of the M grid points in the L-domain. The nine panels, 7a-7i, give the influence of a) $\underline{u}$ observations on $\underline{u}$ corrections, of b) $\underline{v}$ on $\underline{u}$, c) ϕ on $\underline{u}$, d) $\underline{v}$ on $\underline{v}$, e) $\underline{u}$ on $\underline{v}$, f) ϕ on $\underline{v}$, g) ϕ on ϕ, h) $\underline{u}$ on ϕ, i) $\underline{v}$ on ϕ. The particularities of the curves are discussed in the text.

system noise covariance matrix Q: its $\underline{u}$-components were chosen small, cf. (3.12a) and the 4-to-1 ratio of D_1 to D_2. This choice entails relatively good predictions of $\underline{u}$, which have to be corrected only to a small extent by the observations. The (u-u) coefficients (Fig. 7a) are the largest of the coefficients involving $\underline{u}$; they still do not exceed 0.125. The (u-u) influence functions are approximately equal for SF, SL and NY, positive and symmetric in the E-W direction. They are the only ones to have the latter properties.

The influence function for (ϕ-ϕ) centered at SL is the smallest one shown in Fig. 7g. It is positive over land, becoming nearly zero at SF and NY and slightly negative out into the ocean. The relative small size and symmetry of this function is due to its station, SL, being located in the middle of a data-dense region: neighboring stations receive almost equal weights and advection plays but a small role.

The peaks of the (ϕ-ϕ) influence functions centered at NY and at SF are considerably higher than the SL peak. This is due to the absence of observations on the ocean side of these stations. In fact, the peak for the SF influence function is slightly higher than the NY peak. Moreover, the former is located one grid point West of SF, rather than at SF itself, while the NY peak is at NY. Both <u>data density</u> and <u>advection</u> thus play a role.

It makes sense for the point upstream of SF to give even more weight to SF information than SF itself: SF is closer to inland points and their information is also weighted heavily. Due to the advection of error, the forecast error at synoptic time for this ocean point is considerably larger than that for the point downstream from New York, although they are equidistant from land. Hence the larger weight given to adjacent land observations for the Pacific point than for the Atlantic point.

As in Fig. 7g, the (v-v), (v-ϕ) and (ϕ-v) influence functions (Figs. 7d,f,i) all show strong inhomogeneity-differences between the SF, SL and NY functions, as well as anisotropy-differences in the East and the West direction. The SL function for (v-ϕ) and (ϕ-v) is very nearly antisymmetric. This antisymmetry reminds us of the same feature being exhibited by the (v-ϕ) and (ϕ-v) correlations in Schlatter (1975, Fig. 3). The latter were based on assumptions of geostrophy and verified against the network of U. S. radiosonde stations.

Notice from (4.4) that $P_\infty(+)$ and K_∞ have similar symmetry properties, since we took R to be diagonal. The diagonal elements of R^{-1} simply multiply the columns of P_ℓ and $P_{o-\ell}$, yielding the influence functions in Fig. 7. Hence it is legitimate to compare the symmetry properties of the asymptotic influence functions in the case at hand with those of the steady-state covariance matrices in some current objective analysis schemes.

Our $(v-\phi)$ and $(\phi-v)$ influence functions for SF and NY, however, are far from being antisymmetric: they do not even equal zero at SF or NY, respectively. We conclude that it is reasonable to use the geostrophy assumption for wind-height correlations (cf. also Bergman, 1979) in data-dense regions, where the estimation error covariance, i.e., the covariance of forecast-minus-observed fields, is nearly homogeneous and isotropic. Close to the borderline of data-dense and data-poor regions, this assumption will seriously distort the optimal weighting coefficients.

In fact, the influence functions determined by the filtering procedure at the <u>first</u> synoptic time (Fig. 8) are either perfectly symmetric ($(u-u)$, $(v-v)$ and $(\phi-\phi)$) at SL, or perfectly antisymmetric for that station (all six wind-wind and wind-height cross-sections). Furthermore, in all nine panels of Fig. 8, the influence function of NY is either the mirror image ($(u-u)$, $(v-v)$ and $(\phi-\phi)$) of the one of SF, or the inverted mirror image thereof (all other cross-sections).

The comparison of Fig. 8 with Fig. 7 allows us to distinguish between the effect of inhomogeneous data density and the effect of advection on the optimal K-B filter. Fig. 8 shows the effect of data distribution only, since at the first synoptic time no information has been advected yet from previous data insertions. Fig. 7 shows the combination of the two effects.

Different data densities result in different influence functions according to station location (Fig. 8): stations

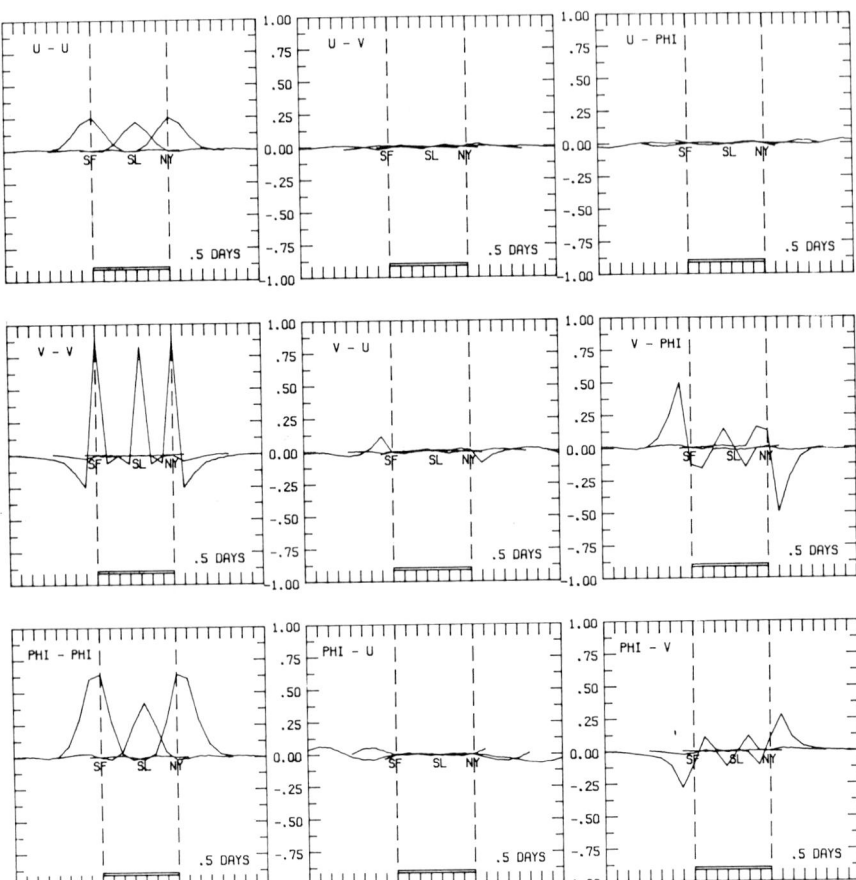

Fig. 8 Same as Fig. 7, but for K used at the first synoptic time. The influence functions here are much more symmetric than in Fig. 7.

located in sharp gradients of observation availability, such
as SF and NY, have more influence than inland stations (SL);
their influence out to sea is also greater than their
influence inland. It is advection, however, which leads
to the difference between the influence functions of
stations on the West Coast (SF) and East Coast of continents
(NY = Tokyo). The latter difference was discussed in connection
with Fig. 7g, and can also be found in Figs. 7d,f,i.

4.2.4 Asymptotic filter: results

We used the gain matrix K at day 10 from this run
as a time-invariant gain matrix for another run which
was otherwise identical to the previous one. This
matrix is a very good approximation to the exact asymptotic
filter K_∞. Estimation errors after 1-2 days were practically
indistinguishable from those obtained when using the
time-varying K-B filter, K_k. There is therefore no need,
in our constant-coefficient system, to compute the filter K
at every synoptic time: the approximate computation of the
asymptotic filter K_∞ once and for all is sufficient for any
practical purposes. Furthermore, K_∞ is independent of P_0. As
indicated in Sec. 4.3, it depends only on Ψ, Q and R.

The asymptotic filter is sometimes called the <u>Wiener
filter</u> (W-filter). Wiener (1949) in fact solved the
estimation problem for stationary time series, using all
past information. It was the contribution of Kalman (1960)
to devise a practical sequential filter for stochastic

processes governed by differential equations and using only past information over a finite time interval.

The individual rms errors in estimation for the K-filter run with system noise, as well as for the W-filter run (not shown), are even noisier than those in Fig. 3. We plotted the ratios of individual m-s errors to the expected m-s error (given in Fig. 6). A comparison of these ratios with Figs. 3b,d showed that the system noise between observations causes fluctuations of higher frequency than did the observational noise alone in the perfect model. The same remarks as in Sec. 4.1 about the actual spread of estimation errors around their mean apply.

4.2.5 Filtering of fast waves. Fig. 9 gives the time histories of the estimated solution at HA, SF and NY. System noise clearly excites the fast, inertia-gravity waves in the solution even more than observational noise alone did in Fig. 4. The time histories for a run with the Π-filter, instead of the K-filter, are shown in Fig. 10. Clearly the fast waves have been removed, and the evolution of the estimated solution is perfectly smooth.

Expected rms errors (not shown) with the Π-filter were, for the v- and ϕ-components, almost indistinguishable from those with the K-filter. Expected rms errors for the u-component, however, were significantly greater in the case of the Π-filter: after 10 days, they were about twice the corresponding errors with the K-filter. This is still well below the level of observational noise, though.

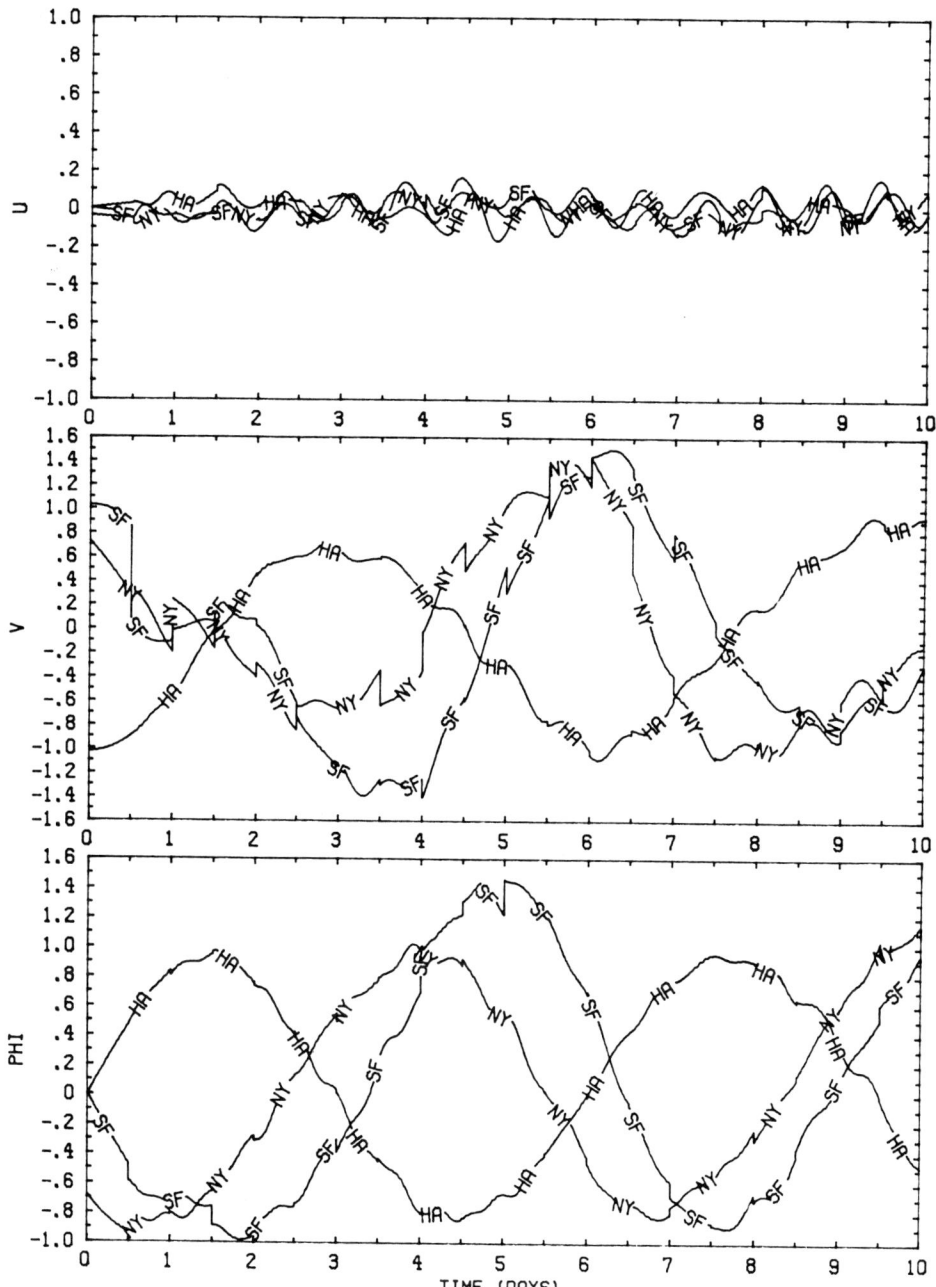

Fig. 9 Same as Fig. 4, but for $Q \neq 0$. The fast oscillations are larger than in the noise free case.

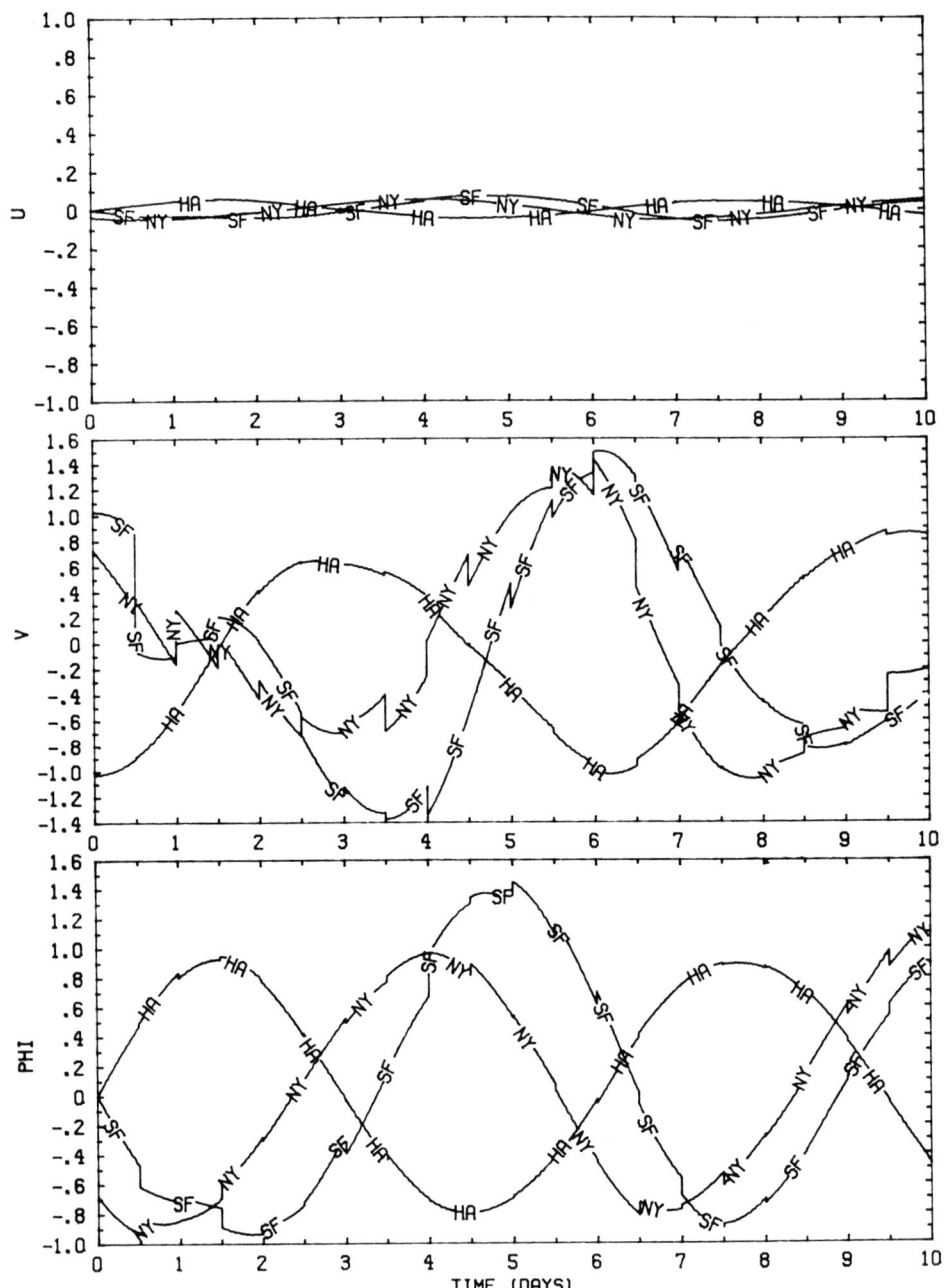

Fig.10 Same as Fig. 9, but for a run using the Π-filter, rather than the K-filter. The fast oscillations have been eliminated.

The larger u-component errors for the Π-filter are explained by the fact that the Π-filter allows almost no observational correction to be performed on the u-components: our slow wave subspace R has very small u-components (cf. Eq. (3.11)). The estimation errors in u produced by the system noise $\underset{\sim}{\zeta}$ cannot be counteracted therefore by the observations.

A run with the asymptotic form of the Π-filter, which is simply ΠK_∞, gave practically the same results as the Π-filter itself.

4.3 Theoretical Analysis of the Scalar Case

In order to help explain some of the qualitative features of the numerical results in Sec. 4.1 and 4.2, we perform an asymptotic analysis of the filtering equations for a <u>scalar</u> state x. The matrices of interest will now actually be scalars, and we assume $\Psi \neq 0$, $H = 1$, $Q \geq 0$, $R > 0$, and $P_0 > 0$. The positivity assumptions on Q, R, and P_0 are due to the fact that they are <u>variances</u>. We assume furthermore that the observation is performed only every $\underline{r}$ time steps; in the numerical experiments reported herein we would have $r = 24$, as $\Delta t = 30$ min. and observations are taken at the standard 12 hour synoptic intervals.

The filtering algorithm (2.11) yields in this case

$$P_k(-) = \Psi^2 P_{k-1}(+) + Q , \qquad (4.5a)$$

$$P_k(+) = \begin{cases} P_k(-)R/(P_k(-) + R) , & \text{when } k=jr, \ j=1,2,3,\ldots, \\ P_k(-) & , \text{otherwise,} \end{cases} \qquad (4.5b)$$

$$K_{jr} = P_{jr}(+)/R. \qquad (4.5c)$$

From Eq. (4.5b), one immediately finds that

$$P_{jr}(+) \leq \min \{P_{jr}(-), R\}, \qquad (4.5d)$$

an analogue of Eq. (2.3b). In particular, the estimation error variance drops below the observational noise level at each observation time, although it may grow in between observation times.

Defining S_j,

$$S_j = P_{jr}(+), \quad j = 0,1,2,\ldots,$$

to be the estimation error variance after the jth observation, and the quantities

$$A = \psi^{2r}, \quad B = \sum_{p=0}^{r-1} \psi^{2p},$$

Eqs. (4.5a,b) can be converted into a nonlinear difference equation for S_j:

$$S_j = g(S_{j-1}), \qquad (4.6a)$$

where

$$g(s) = \frac{(As + BQ)R}{As + BQ + R}. \qquad (4.6b)$$

To determine the asymptotic behavior of Eq. (4.6), and hence of the filter K, we now distinguish between the two cases $Q = 0$ and $Q > 0$. In the case of a perfect model, $Q = 0$, the solution of (4.6) may be written explicitly as

$$S_j = \frac{A^j(A-1)S_0 R}{A(A^j-1)S_0 + (A-1)R}, \quad \text{if } |\psi| \neq 1, \qquad (4.7a)$$

or

$$S_j = \frac{S_0 R}{jS_0 + R}, \quad \text{if } |\psi| = 1. \qquad (4.7b)$$

As $j \to +\infty$, therefore,

$$S_j \to 0, \quad \text{if } |\psi| \leq 1, \qquad (4.7c)$$

$$S_j \to (1 - \frac{1}{A})R, \qquad \text{if } |\Psi| > 1. \qquad (4.7d)$$

Eq. (4.7c) states that for <u>stable</u> dynamics, i.e., $|\Psi| \leq 1$, the estimation error variance, and hence the filter $K_{jr} = S_j/R$, always tend to zero. System (3.1) itself is conservative, while our difference scheme (3.6) is dissipative. Hence all eigenvalues of the difference scheme matrix Ψ have modulus less than or equal to unity. We are thus in the stable case (4.7c) and it is only to be expected that our estimation error covariance matrices and filter K_k approach zero in the absence of system noise. This is in accordance with our discussion of Fig. 2 in Sec. 4.1.

We wish to determine now the asymptotic nature of S_j in the case $Q > 0$, i.e., in case system noise is present. The quadratic equation

$$s = g(s) \qquad (4.8)$$

has a positive discriminant, hence it has real roots. Its free term is negative, hence the roots are of opposite sign. Let S_+ denote the unique positive root of (4.8).

Notice that

$$dg/ds = AR^2/(As+BQ+R)^2 > 0.$$

Therefore $g(s)$ is a monotone increasing function of s, with $g(0) > 0$, while its derivative is monotone decreasing and tending to zero as $s \to +\infty$. It follows that the root S_+ of (4.8) is approached monotonically by the solutions of the recursion (4.6a) (Isaacson and Keller, 1966, Ch. 3.1, Fig. 2a),

$$S_j \to S_+ \quad \text{as} \quad j \to +\infty .$$

If S_0 is greater than (less than) S_+, then S_j will decrease (increase) monotonically to S_+. Since $K_{jr} = S_j/R$, we have also

$$K_{jr} \to S_+/R \quad \text{as} \quad j \to +\infty,$$

the convergence being monotone as well. This is in accordance with the monotone decrease of trace $P_{jr}(+)$ in Fig. 6; in fact $P_{jr}(-)$ decreases also monotonically. Furthermore, S_+ is independent of $S_0 = P_0$, and hence the asymptotic filter $K_\infty = S_+/R$ is independent of P_0 as well.

To find an approximate value for S_+, we now assume that $|\Psi| < 1$ and $r \gg 1$, so that $A = \Psi^{2r} \ll 1$. Then the quadratic term in Eq. (4.8) is negligible and we have, approximately,

$$S_+ = \frac{QR}{Q + (1-\Psi^2)R} . \qquad (4.9a)$$

It follows in general, by analogy with (2.3b) and (4.5d), that

$$S_+ \leq \min \{R, Q/(1-\Psi^2)\} ,$$

at least approximately. In particular, when the observational error variance R is small enough, so that

$$(1-\psi^2)R \ll Q , \qquad (4.9b)$$

then S_+ is roughly equal to R, and the size of Q has little influence. If, on the other hand, the observational error variance is large enough so that

$$Q \ll (1 - \psi^2)R , \qquad (4.9c)$$

then S_+ is roughly proportional to Q, and the size of R has little influence.

The two extreme cases (4.9b,c) explain much of the qualitative nature of the results in Sec. 4.2. Indeed, a matrix version of Eq. (4.9b) is satisfied over land, and we see that the expected m-s errors at observation times are approximately equal to the observational error variances. In fact, they are slightly smaller than R; this is in accordance with (4.5d), as well as (4.9a,b). It also agrees with operational experience, as stated in Sec. 3.4.

Over the ocean, we can write $R = \infty$, so that a matrix version of (4.9c) is satisfied. Experiments with different magnitudes of Q (not shown), have confirmed that the size of Q is indeed the determining factor in the size of P over the ocean. This also agrees with operational experience: analysis error in data-poor regions is essentially equal to the error in forecasts from one analysis time (synoptic or subsynoptic) to the next.

The asymptotic properties of P and K discussed above have counterparts in the full, vector-matrix case, under the assumptions of <u>complete observability</u> and <u>complete controllability</u>. These assumptions concern properties of the matrices Ψ and H; they are satisfied for our model. The interested reader is referred to Bucy and Joseph (1968, Ch. 5) and to Jazwinski (1970, Sec. 7.6) for a full discussion of the general results.

5. CONCLUDING REMARKS

We have shown the ways in which the concepts and formalism of sequential estimation theory are relevant to the 4-D assimilation of meteorological data, by applying them to a simple model. The stochastic-dynamic model used for the illustration of the theory was governed by the linear shallow-water equations, including the β-effect of latitudinal changes in planetary vorticity. The dynamics of this model are similar to those of operational NWP models in that they admit as solutions slow, quasi-geostrophic Rossby waves, as well as fast inertia-gravity waves.

We have modified the standard Kalman-Bucy (K-B) filter in order to obtain optimal estimates of the slow, meteorologically significant waves, while eliminating entirely the fast, undesirable waves. In this way, our modified K-B filter achieves simultaneously the optimal 4-D assimilation of data and the initialization of model states for the purpose of noise-free forecasts.

It was shown that the optimal filter for the linear problem converges rapidly to an asymptotic matrix, the Wiener filter. Furthermore, the asymptotic filter (W-filter) performs nearly as well as the exact, time-varying filter (K-B filter). The Wiener filter depends on system dynamics Ψ, observational pattern H, system noise covariance Q and observational noise covariance R; it does not depend on the initial errors, P_0. The rapid convergence and good performance of the W-filter for the linear problem hold hope for the filtering of nonlinear problems with similar dynamic and stochastic properties.

Nonlinear estimation theory is not completely understood mathematically, at least not in a practically applicable form. The filter which is most widely used in engineering applications is the extended K-B filter (EKF: Bucy and Joseph, 1968, Ch. 8; Gelb, 1974, Ch. 6; Jazwinski, 1970, Ch. 6 and Ch. 9). The principle of the EKF is simple: linearize the problem around an estimated state, and apply the corresponding linear filter over a time interval, T say, over which the solution of the nonlinear problem is not expected to change much. In large-scale NWP this time could equal 6 h to 24 h. After time T, relinearize around the new state and proceed. Clearly, T is limited by the dynamics of the system in the problem. When choosing T, a trade-off between accuracy and expediency has to be made.

The EKF gives good results when the true characteristic correlation time τ of the perturbations, which we model as white noise, i.e., $\tau = 0$, is actually short compared to T, $\tau \ll T$. This is certainly the case in NWP. Moreover, it appears from our experience with linear problems that:
a) the asymptotic filter for each one of the successive linearizations will work sufficiently well, and there is no need to compute time-varying filters over a time interval T;
b) the dependence of the W-filter on (linear) system dynamics Ψ is rather weak.

We conclude that the W-filter for the succeeding T-interval will be easily computable from the W-filter valid over the preceding T-interval by a perturbation procedure.

It is more realistic in NWP to let the observation matrix H be a function of time, $H = H(t)$, rather than a constant. Different observations are made at the synoptic times, 0000 GMT and 1200 GMT, the subsynoptic times, 0600 GMT and 1800 GMT, and in between. The distribution of observations, however, is not too far from being time-periodic. Let $H^*(t)$ be a periodic matrix function of time, with period 24 h, and assume that the actual observation matrix, $H(t)$, differs from $H^*(t)$ at every t only by a matrix of small rank, i.e., only by the presence, absence or location of relatively few observations.

The asymptotic filter corresponding to $\bigl(\Psi(t), H^*(t)\bigr)$, $K^*(t)$ say, should also be periodic, with a period of 24 h, rather than constant. We expect some easily computed modification of $K^*(t)$ to be a good approximation to the optimal filter for $\bigl(\Psi(t), H(t)\bigr)$. We plan to study, therefore, time-periodic <u>observation patterns</u> and their modifications.

Sequential estimation accounts explicitly for the fact that the system whose state we wish to estimate is governed by certain dynamics. It is this aspect of the theory which distinguishes it from the so-called "optimal interpolation" currently used in operational NWP. The latter essentially assumes that the system obeys trivial dynamics, $\Psi = I$.

We saw already that taking into account the dynamics allowed us to unify the assimilation and intialization aspects of preparing initial data for numerical weather forecasts. Furthermore, it allows us to account in a systematic way for the <u>advection of information</u> from data-rich to data-poor areas, and of "negative information", i.e., large errors, from data-poor to data-rich regions.

In particular, the theory shows that weighting coefficients for observations should be skewed in the direction of the prevailing winds, with larger weights upstream; the amount of skewness should depend on average wind intensity, i.e., on season. Also weights used on the Western edge of the continents should be different from those used on the Eastern edge. The present results on the anisotropy and inhomogeneity of estimation error structure in the zonal direction should also be supplemented by the results on inhomogeneity in the latitudinal direction of Ghil <u>et al</u>. (1980). It is this aspect of the theory which we expect to have the largest impact in terms of improving operational procedures.

One further aspect of the theory merits attention: the covariance matrices Q and R do not need to be prescribed <u>a priori</u>. They can be determined in the estimation process itself (e.g., Chin, 1979; Ohap and Stubberud, 1976) by using an <u>adaptive</u> filter. The determination of system noise would have important consequences for predictability theory, as well as for stochastic modifications of numerical schemes (Faller and Schemm, 1977). The determination of observational noise, eliminating the well known problem of "ground truth", would also help greatly in improving operational objective analysis and data assimilation.

Acknowledgement.

The potential usefulness of estimation theory in meteorological applications was first pointed out to one of us (M.G.) by Edward S. Sarachik. It is a pleasure to acknowledge many useful discussions on linear and nonlinear filtering with George Papanicolaou. Practical aspects of the filters' eventual implementation in an operational environment were discussed with Wayman Baker, Mark Cane, Milton Halem and Eugenia Kalnay-Rivas. The final version of the manuscript benefited from comments by G. Cats, R. Daley, A. Hollingsworth, A. Lorenc and O. Talagrand, made during and after the Seminar. The manuscript could not have been prepared without the whole-hearted support of Constance Engle.

Our work on this problem has been supported by NASA Grants NSG-5034 and NSG-5130, administered by the Laboratory for Atmospheric Sciences, Goddard Space Flight Center. Computations were carried out on the CDC 6600 of the Courant Mathematics and Computing Laboratory, New York University, under Contract DE-AC02-76ER03077 with the U. S. Department of Energy.

References

Bengtsson, L., 1975: 4-Dimensional Assimilation of Meteorological Observations, GARP Publications Series, No. 15, World Meteorological Organization — International Council of Scientific Unions, CH-1211 Geneva 20, Switzerland, 76 pp.

Bergman, K. H., 1979: Multivariate analysis of temperatures and winds using optimum interpolation, Mon. Wea. Rev. 107, 1423-1444.

Browning, G., A. Kasahara and H.-O. Kreiss, 1979: Initialization of the primitive equations by the bounded derivative method, NCAR Ms. 0501-79-4, National Center for Atmospheric Research, Boulder, Colo. 80307, 44 pp.

Bube, K. P., and M. Ghil, 1980: Assimilation of asynoptic data and the initialization problem, this volume.

Bucy, R. S., and P. D. Joseph, 1968: Filtering for Stochastic Processes with Applications to Guidance, Wiley-Interscience, New York, 195 pp.

Charney, J., M. Halem and R. Jastrow, 1969: Use of incomplete historical data to infer the present state of the atmosphere, J. Atmos. Sci. 26, 1160-1163.

Chin, L., 1979: Advances in adaptive filtering, in Control and Dynamic Systems, Vol. 15, C. T. Leondes (ed.), Academic Press, 278-356.

Curtain, R. F., and A. J. Pritchard, 1978: Infinite Dimensional Linear Systems Theory, Lecture Notes in Control and Information Sciences, Vol. 8, Springer-Verlag, New York, 297 pp.

Davis, M.H.A., 1977: Linear Estimation and Stochastic Control, Halsted Press, John Wiley and Sons, New York, 224 pp.

Faller, A. J., and C. E. Schemm, 1977: Statistical corrections to numerical prediction equations II, <u>Mon. Wea. Rev. 105</u>, 37-56.

Fleming, R. J., T. M. Kaneshige and W. E. McGovern, 1979a: The global weather experiment I. The observational phase through the first special observing period, <u>Bull. Amer. Met. Soc. 60</u>, 649-659.

_____, _____, _____, and T. E. Bryan, 1979b: The global weather experiment II. The second special observing period, <u>Bull. Amer. Met. Soc. 60</u>, 1316-1322.

Gelb, A. (ed.), 1974: <u>Applied Optimal Estimation</u>, The M.I.T. Press, Cambridge, Mass., 374 pp.

Ghil, M., 1980: The compatible balancing approach to initialization, and four-dimensional data assimilation, <u>Tellus 32</u>, 198-206.

_____, and R. Mosebach, 1978: Asynoptic variational method for satellite data assimilation, in Halem <u>et al</u>. (1978), pp. 3.32-3.49.

_____, R. C. Balgovind and E. Kalnay-Rivas, 1980: <u>A stochastic-dynamic model for global atmospheric mass field statistics</u>, NASA Tech. Memo. 82009, NASA Goddard Space Flight Center, Greenbelt, Md. 20771, 50 pp.

_____, M. Halem and R. Atlas, 1979: Time-continuous assimilation of remote-sounding data and its effect on weather forecasting, <u>Mon. Wea. Rev. 107</u>, 140-171.

Halem, M., M. Ghil, R. Atlas, J. Susskind and W. J. Quirk, 1978: The GISS sounding temperature impact test. NASA Tech. Memo. 78063, NASA Goddard Space Flight Center, Greenbelt, Md. 20771, 421 pp. [NTIS N7831667].

Isaacson, E., and H. B. Keller, 1966: Analysis of Numerical Methods, Wiley, New York, 541 pp.

Jazwinski, A. H., 1970: Stochastic Processes and Filtering Theory, Academic Press, New York, 376 pp.

Jones, R. H., 1965a: Optimal estimation of initial conditions for numerical prediction, J. Atmos. Sci. 22, 658-663.

_____, 1965b: An experiment in nonlinear prediction, J. Appl. Meteor. 4, 701-705.

Kalman, R. E., 1960: A new approach to linear filtering and prediction problems, Trans. ASME, Ser. D, J. Basic Eng., 82, 35-45.

_____, and R. S. Bucy, 1961: New results in linear filtering and prediction theory, Trans. ASME, Ser. D.: J. Basic Eng., 83, 95-108.

Leith, C. E., 1978: Objective methods for weather prediction, Ann. Rev. Fluid Mech., 10, 107-128.

_____, 1980: Nonlinear normal mode initialization and quasi-geostrophic theory, J. Atmos. Sci., 37, 958-968.

Lorenz, E. N., 1969: The predictability of a flow which possesses many scales of motion, Tellus, 21, 289-307.

McPherson, R. D., K. H. Bergman, R. E. Kistler, G. E. Rasch and D. S. Gordon, 1979: The NMC operational global data assimilation system, Mon. Wea. Rev., 107, 1445-1461.

Miyakoda, K., and O. Talagrand, 1971: The assimilation of past data in dynamical analysis. I, Tellus, 23, 310-317.

Ohap, R. F., and A. R. Stubberud, 1976: Adaptive minimum variance estimation in discrete-time linear systems, in Control and Dynamic Systems, Vol. 12, C. T. Leondes (ed.), Academic Press, 583-624.

Palmén, E., and C. W. Newton, 1969: Atmospheric Circulation Systems, Academic Press, New York, 603 pp.

Parzen, E., 1960: Modern Probability Theory and Its Applications, Wiley, New York, 464 pp.

Pedlosky, J., 1979: Geophysical Fluid Dynamics, Springer-Verlag, New York, 624 pp.

Petersen, D. P., 1968: On the concept and implementation of sequential analysis for linear random fields, Tellus, 20, 673-686.

_____, 1970: Algorithms for sequential and random observations, Meteorol. Mono., 11, 100-109.

_____, 1973a: Transient suppression in optimal sequential analysis, J. Appl. Meteor., 12, 437-440.

_____, 1973b: A comparison of the performance of quasi-optimal and conventional objective analysis schemes, J. Appl. Meteor, 12, 1093-1101.

_____, 1973c: Static and dynamic constraints on the estimation of space-time covariance and wavenumber-frequency spectral fields, J. Atmos. Sci., 30, 1252-1266.

_____, 1976: Linear sequential coding of random space-time fields, Info. Sci., 10, 217-241.

Phillips, N. A., 1971: Ability of the Tadjbakhsh method to assimilate temperature data in a meteorological system, J. Atmos. Sci., 28, 1325-1328.

_____, 1976: The impact of synoptic observing and analysis systems on flow pattern forecasts, Bull. Amer. Met. Soc., 57, 1225-1240.

Richtmyer, R. D., and K. W. Morton, 1967: Difference Methods for Initial-Value Problems, 2nd. ed., Wiley-Interscience, New York, 405 pp.

Rutherford, I. D., 1972: Data assimilation by statistical interpolation of forecast error fields, J. Atmos. Sci., 29, 809-815.

Schlatter, T. W., 1975: Some experiments with a multivariate stiatistical objective analysis scheme, Mon. Wea. Rev., 103, 246-257.

_____, G. W. Branstator and L. G. Thiel, 1976: Testing a global multivariate statistical objective analysis scheme with observed data, Mon. Wea. Rev., 104, 765-783.

_____, 1977: Reply (to Comments by H. J. Thiebaux on "Testing a global ..."), Mon. Wea. Rev., 105, 1465-1468.

Tadjbakhsh, I. G., 1969: Utilization of time-dependent data in running solution of initial value problems, J. Appl. Meteor., 8, 389-391.

Wiener, N., 1949: Extrapolation, Interpolation and Smoothing of Stationary Time Series with Engineering Applications, Wiley, New York, 163 pp.

Appendix A. List of Major Symbols

$c_\ell^{(m)}$ for solutions of the continuous system (3.1), the three phase speeds, m = 1,2,3, corresponding to wave number ℓ, as defined by Eq. (3.3b). The same notation is used for the phase speeds of solutions of the discrete system (3.6), which agree with the phase speeds of solutions of the continuous system to order $(\Delta t)^2$.

E   expectation, or ensemble averaging, operator

f   Coriolis parameter, $f = 10^{-4} s^{-1} \cong 2\Omega \sin 45°$

G   fast wave subspace

g   gravitational acceleration constant of the Earth

H   observation matrix; defines the observed linear combinations of state variables

j   superscript indicating the spatial grid point, $x^j = j\Delta x$

K   Kalman gain matrix, defined by (2.11c)

k   subscript indicating the time, $t_k = k \Delta t$

L   length of the computational domain, about half the circumference of the Earth at 45°N

ℓ wave number; $\ell L/2\pi$ is an integer

M   number of grid points in the L-domain

n   number of state variables, n = 3M; the dimension of $\underset{\sim}{x}$

$P_k(-)$ estimation error covariance matrix just prior to observations at time k

$P_k(+)$ estimation error covariance matrix just after observations at time k

p	number of observations at a synoptic time; the dimension of $\underset{\sim}{z}$
Q	covariance matrix of the system noise
$\mathcal{R}$	slow wave subspace
R	covariance matrix of the observational noise
t	time
Δt	temporal increment in the discrete system (3.6)
U	mean zonal wind component in the linearization of (3.2)
u	perturbation zonal wind component
v	perturbation meridional wind component
v_{max}	initial amplitude of v
$\underset{\sim}{W}$	a solution $(u,v,\phi)^T$ of the continuous system (3.1)
$\underset{\sim}{W}_k^j$	a solution (u_k^j, v_k^j, ϕ_k^j) of the discrete system (3.6)
x	distance along the spatial L-domain
Δx	spatial increment in the discrete system (3.6)
$\underset{\sim}{x}$	"true" atmospheric state given by the stochastic model (2.5)
$\hat{\underset{\sim}{x}}_k(-)$	estimated atmospheric state just prior to observations at time k
$\hat{\underset{\sim}{x}}_k(+)$	estimated atmospheric state just after observations at time k
$\underset{\sim}{z}$	vector of observations, given by (2.6)
$\underset{\sim}{\zeta}$	observational noise, or random part of the observation model (2.6)
$\underset{\sim}{\xi}$	system noise, or random part of the atmospheric model (2.5)
Π	orthogonal projection operator onto the slow wave subspace $\mathcal{R}$

Φ	mean geopotential at 45°N in the linearization of (3.2), i.e., g times the mean equivalent atmospheric height at 45°N
ϕ	perturbation geopotential
ϕ_0	initial amplitude of ϕ
Ψ	matrix defining the atmospheric dynamics, given for our model by Eqs. (3.6)
Ω	angular rate of rotation of the Earth

CONVERGENCE OF ASSIMILATION PROCEDURES

Olivier Talagrand

ECMWF*

These notes deal with a number of mathematical problems related to four-dimensional data assimilation. They are based on a general criterion for convergence of an assimilation procedure, which is derived in Sections 1 and 2, and then applied to various cases.

The basis for the content of these notes has been described in detail in a previous publication (Talagrand, 1977) which will be hereafter referred to as T77.

1. MATHEMATICAL PRELIMINARIES

We will formalise the assimilation problem by assuming that we observe the time evolution of a physical system (which will in applications be the atmosphere, but no specific hypothesis is necessary at this stage). The system will be assumed to be made of two parts: a first part X (e.g. the mass field) which is observed at successive instants, and whose state at any time is defined by the values of a finite number p of independent parameters; and a second part Y (e.g. the wind field) which is to be reconstituted through an assimilation process, and whose state at any time is defined by the values of q independent parameters.

The time evolution of the system is supposed to be described by a set of $r = p+q$ differential equations of first order with respect to time, which can be summed up in the two vector equations

$$\frac{dX}{dt} = F[X,Y] \qquad (1a)$$

$$\frac{dY}{dt} = G[X,Y] \qquad (1b)$$

where F and G are respectively a p-valued and a q-valued function of X and Y. F and G could be assumed to depend also on time, without this resulting in any modification to the sequel. Equations (1) can be thought of as being, for instance, the meteorological primitive equations.

*Present affiliation: Laboratoire de Meteorologie Dynamique du CNRS-ENS Physique, Paris, France

An arbitrary initial time t_o being chosen, any initial conditions $[X_o, Y_o]$ at t_o define a unique solution to Equations (1). The values assumed by that solution at time t will be noted $[X(X_o,Y_o;t), Y(X_o,Y_o;t)]$. Any initial perturbations $[\Delta X_o, \Delta Y_o]$ on $[X_o,Y_o]$ define a solution

$$[X(X_o,Y_o;t) + \Delta X(t), Y(X_o,Y_o;t) + \Delta Y(t)],$$

where the time evolution of ΔX and ΔY is described by the <u>perturbation equations</u> in the vicinity of the solution $[X(X_o,Y_o;t), Y(X_o,Y_o;t)]$

$$\frac{d\Delta X}{dt} = F[X(X_o,Y_o;t) + \Delta X, Y(X_o,Y_o;t) + \Delta Y] - F[X(X_o,Y_o;t), Y(X_o,Y_o;t)] \quad (2a)$$

$$\frac{d\Delta Y}{dt} = G[X(X_o,Y_o;t) + \Delta X, Y(X_o,Y_o;t) + \Delta Y] - G[X(X_o,Y_o;t), Y(X_o,Y_o;t)] \quad (2b)$$

These equations are readily obtained from (1). Linearising them with respect to ΔX and ΔY (and changing ΔX and ΔY to δX and δY in order to avoid any confusion with (2)) leads to

$$\frac{d\delta X}{dt} = \frac{DF}{DX}(t)\,\delta X + \frac{DF}{DY}(t)\,\delta Y \quad (3a)$$

$$\frac{d\delta Y}{dt} = \frac{DG}{DX}(t)\,\delta X + \frac{DG}{DY}(t)\,\delta Y \quad (3b)$$

where $\frac{DF}{DX}(t)$ is the p x p jacobian matrix made up of the partial derivatives of F with respect to the components of X. The argument t means that these derivatives are taken at point $[X(X_o,Y_o;t), Y(X_o,Y_o;t)]$. $\frac{DF}{DY}(t)$, $\frac{DG}{DX}(t)$, $\frac{DG}{DY}(t)$ are similarly defined jacobian matrices, with respective dimensions p x q, q x p and q x q. Equations (3) are the <u>linearised perturbation equations</u> in the vicinity of the solution $[X(X_o,Y_o;t), Y(X_o,Y_o;t)]$. Unless the basic equations (1) are themselves linear, there is one different such set of linearised equations for every solution of (1).

The solution of equations (3) at any time t depends linearly on the initial conditions $[\delta X_o, \delta Y_o]$ at time t_o. This is expressed by the following equations

$$\delta X(t) = R_x^x(t,t_o)\,\delta X_o + R_x^y(t,t_o)\,\delta Y_o \quad (4a)$$

$$\delta Y(t) = R_y^x(t,t_o)\,\delta X_o + R_y^y(t,t_o)\,\delta Y_o \quad (4b)$$

where $R_x^x(t,t_o)$, $R_x^y(t,t_o)$, $R_y^x(t,t_o)$, $R_y^y(t,t_o)$ are four matrices with respective dimensions p x p, p x q, q x p, q x q, which together make up a square matrix $R(t,t_o)$ of order r

$$R(t,t_o) = \begin{pmatrix} R_x^x(t,t_o) & R_x^y(t,t_o) \\ \\ R_y^x(t,t_o) & R_y^y(t,t_o) \end{pmatrix} \qquad (5)$$

$R(t,t_o)$ is called the <u>resolvent matrix</u> of the linearised equation (3) between times t_o and t.

We will be interested in the linearised perturbation equation (3) for the following reason: for "small" initial perturbations $[\Delta X_o, \Delta Y_o]$, the time evolution of the resulting perturbation $[\Delta X(t), \Delta Y(t)]$ is described to some approximation by equation (3). This vague statement is made precise by the following theorem (for a proof, see e.g. Coddington and Levinson (1955)).

<u>Theorem (P)</u> <u>Given initial perturbations $[\Delta X_o, \Delta Y_o]$ at time t_o, the difference at any given time t between the corresponding solutions of the exact and linearised sets of perturbation equations (2) and (3) is an infinitesimal of higher order than $[\Delta X_o, \Delta Y_o]$.</u>

This can be expressed by the following equations

$$\Delta X(t) = R_x^x(t,t_o)\Delta X_o + R_x^y(t,t_o)\Delta Y_o + o(\Delta X_o, \Delta Y_o) \qquad (6a)$$

$$\Delta Y(t) = R_y^x(t,t_o)\Delta X_o + R_y^y(t,t_o)\Delta Y_o + o(\Delta X_o, \Delta Y_o) \qquad (6b)$$

where, following a standard notation, $o(\Delta X_o, \Delta Y_o)$ denotes an infinitesimal of higher order than $[\Delta X_o, \Delta Y_o]$, i.e. a function of $[\Delta X_o, \Delta Y_o]$ such that the ratio $\dfrac{|o(\Delta X_o, \Delta Y_o)|}{|\Delta X_o| + |\Delta Y_o|}$ tends to 0 when ΔX_o and ΔY_o tend to 0.

Theorem (P) essentially means that the values $[X(X_o, Y_o; t), Y(X_o, Y_o; t)]$ assumed at time t by a solution of the basic equation (1) are continuous and differentiable functions of the initial conditions $[X_o, Y_o]$, and that the corresponding partial derivatives are the entries of the resolvent matrix $R(t,t_o)$ of (5).

Some further properties of the resolvent matrix will be useful. Since the choice of the initial time t_o is arbitrary, $R(t',t'')$ is defined for any two times t' and t'' at which the solution $[X(X_o, Y_o; t), Y(X_o, Y_o; t)]$ is itself defined, whether $t' \geq t''$ or $t'' \geq t'$. Since integrating equation (3) from t' to t'', and then to a third time t''' will produce the same result as integrating directly from t' to t''', the corresponding resolvent matrices must satisfy the following relationship

$$R(t''', t'') \, R(t'',t') = R(t''',t') \qquad (7)$$

which, when t''' = t', reduces to

$$R(t',t'') \, R(t'',t') = I \qquad (8)$$

where I is the unit matrix of order r.

Taking into account decomposition (5), each of the two above relationships can be readily transformed into four relationships between the corresponding matrices R_x^x, R_x^y....

2. CONVERGENCE OF AN ASSIMILATION PROCEDURE

For the sake of definiteness, we will make the following two hypotheses about the assimilation process

a) The part X has been observed at N successive instants $t_1, t_2, \ldots, t_N$ and the assimilation is performed according to the following forward-backward procedure: the evolution equations of the system are integrated alternatively forward and backward in time over the time period $[t_1, t_N]$. Whenever the integration time reaches an observation time t_i, whether in a forward or in a backward integration, the values predicted for X are replaced with the observed values, while the values predicted for Y are not modified. The model integration is then resumed, and carried to the next observation time, at which a new updating is performed.

This procedure calls for an important remark. The nature of the parameters which make up X is imposed by the observations, but the nature of the parameters which make up Y is not imposed by the conditions of the problem. The nature of Y can be chosen arbitrarily under the only condition that X and Y together uniquely define the complete state of the system under observation. For instance, in the case of an atmospheric model, with X representing, say, the mass field, one can choose for Y either the velocity field $\mathbf{V}$, or the momentum field $\rho\mathbf{V}$, or any other combination of the mass and velocity field which, together with the mass field, completely defines the state of the flow. These different choices are obviously not equivalent for the assimilation. The nature of X being given, there are infinitely many possible choices for the nature of Y. For this reason, the hypothesis that Y is not modified at an introduction time

is much less restrictive than could *a priori* seem. We will come back later
to this point and will assume for the time being that one particular arbitrary
choice has been made for Y.

b) The assimilation is carried out with a numerical model which solves exactly
the evolution equations (1). Moreover, the observations are perfect, which means
they are exact values taken from one solution of equations (1). This solution will
be called the observed solution. The hypothesis made here is the classical "identical twin" hypothesis. We will see later how the results to be derived below are
modified when this hypothesis is relaxed.

2.1 The amplification matrix over an assimilation cycle

We shall choose the latest observation time t_N as the arbitrary origin of the
successive forward-backward assimilation cycles. Setting $t_N - t_1 = T$, it will
be convenient to introduce along each assimilation cycle an auxiliary time
variable τ, defined modulo $2T$. This variable will increase from 0 to T in
the backward phase of the cycle, and from T to 2T in the forward phase. To
each of its values τ, there corresponds a unique value of the physical time. To
any value t of the physical time ($t_1 < t < t_N$) correspond two values,
τ and τ' of the auxiliary time (such that $\tau + \tau' = 2T$) belonging respectively to the backward and forward phases of the assimilation cycle. The instants
in the cycle when observations of X are introduced into the integration procedure
will be denoted τ_1 (coinciding with t_N), $\tau_2, \ldots, \tau_N$ (coinciding with t_1),
$\tau_{N+1}, \ldots, \tau_{2(N-1)}$ (coinciding with t_{N-1}). We shall define $M = 2(N-1)$.
M introductions of observations are performed in the course of one assimilation
cycle.

For any two values τ and τ' of the auxiliary time variable, it will be
convenient to denote by $R(\tau',\tau)$ the resolvent matrix (5) between the corresponding values of the physical time. A similar notation will be used for the
submatrices R^x_x, R^y_x, ...

Let us now consider the difference $(\Delta X, \Delta Y)$ between the assimilating model and
the observed solution. Between two observation times, this difference varies
according to the perturbation Eqs. (2) where the "unperturbed" solution
$[X(X_o,Y_o; t), Y(X_o,Y_o; t)]$ is now the observed solution $[X(t), Y(t)]$. At
an observation time τ_i, ΔX is set equal to 0, while ΔY remains unchanged.
Let ΔY_n be the Y-difference at the end of the n-th assimilation cycle. In
the (n+1)-st cycle, starting at time τ_1, the model integration from τ_1 to
τ_2 will produce a difference ΔX which will be then set equal to 0, and a
difference $\Delta Y(\tau_2)$ which, according to (6b) is equal to

$$\Delta Y(\tau_2) = R_y^y(\tau_2,\tau_1)\,\Delta Y_n + o(\Delta Y_n)$$

Similarly, the subsequent integration of the model from τ_2 to τ_3 will produce a difference

$$\Delta Y(\tau_3) = R_y^y(\tau_3,\tau_2)\,\Delta Y(\tau_2) + o(\Delta Y(\tau_2))$$

$$= R_y^y(\tau_3,\tau_2)\,R_y^y(\tau_2,\tau_1)\,\Delta Y_n + o(\Delta Y_n)$$

This argument, carried out over the complete assimilation cycle, shows that the difference ΔY_{n+1} at the end of the (n+1)-st cycle will be

$$\Delta Y_{n+1} = A\,\Delta Y_n + o(\Delta Y_n) \tag{9}$$

where A is the matrix

$$A = R_y^y(\tau_1,\tau_M)\,R_y^y(\tau_M,\tau_{M-1})\ldots\ldots R_y^y(\tau_2,\tau_1) \tag{10}$$

A is the product of M square matrices of order q each of which represents the effect of the assimilation over one interval (τ_i, τ_{i+1}). A will be called the <u>amplification matrix</u> of the difference ΔY over one assimilation cycle. It is entirely determined by the linearised perturbation Eqs. (3) in the vicinity of the observation solution and, more precisely, by only one, namely R_y^y, of the four submatrices which make up R (see Eq. (5)). It must be noted that, contrary to R, R_y^y does not satisfy a "contraction" relationship of type (7), so that expression (10) cannot be written in a more concise form.

Now, according to a general result of matrix algebra, the relevant parameter for the behaviour of ΔY_n as n tends to infinity is the <u>spectral radius</u> $\rho(A)$, which is by definition the largest modulus of the eigenvalues of A (see e.g. Varga (1962):

- if $\rho(A)$ is strictly less than 1, ΔY_n will tend to 0 as n increases to infinity, provided the initial difference ΔY_o is small enough so that the $o(\Delta Y_o)$ term in (9) is negligible compared with $A\Delta Y_o$.

- if $\rho(A)$ is larger than 1, the components of ΔY_n along eigenvectors corresponding to eigenvalue(s) with modulus larger than 1, will be amplified by the assimilation, and ΔY_n will not tend to 0.

- finally, if $\rho(A)$ is exactly 1, the component of ΔY_n along eigenvectors corresponding to eigenvalue(s) with modulus equal to 1, will be neither amplified nor reduced in the product $A\Delta Y_n$, and the behaviour of ΔY_n as n tends to infinity will depend on the higher order term $o(\Delta Y_n)$ in (9).

We see that $\rho(A) \leq 1$ is a necessary condition for convergence of an assimilation, and $\rho(A) < 1$ a sufficient condition, when convergence is defined as follows: there exists some number $\varepsilon > 0$ such that ΔY_n will tend to 0 as n tends to infinity provided the initial difference ΔY_o is less than ε.

For the sake of simplicity, we have assumed that the "same" p parameters, making up the vector X, are observed at the successive times $t_1, t_2, \ldots, t_N$. This assumption is in effect not necessary and an amplification matrix of type (10) can be obtained through a similar derivation when the nature, or even the number of the observed parameters varies with the observation time t_i.

Also, we have considered only the case of a forward-backward assimilation which, being an exactly iterative process lends itself more easily to a rigorous mathematical treatment. At the price of a somewhat more complicated mathematical formalism, the approach presented above can be extended to a forward assimilation, in which the assimilating model is constantly integrated forward in time, while new data are successively introduced. In this case too, the convergence of the assimilation depends on submatrices R_y^y extracted from the resolvent matrix of the linearised perturbation Eqs. (3). We will see in the next subsection an example of convergence of a purely forward assimilation.

2.2 Application to the linearised shallow-water equations

The non-linear shallow-water equations read

$$\frac{\partial \phi}{\partial t} + \nabla \cdot (\phi \mathbb{V}) = 0 \tag{11a}$$

$$\frac{\partial \mathbb{V}}{\partial t} + (\mathbb{V} \cdot \nabla) \mathbb{V} + \nabla \phi + \mathbb{n} \times f \mathbb{V} = 0 \tag{11b}$$

where ϕ and $\mathbb{V}$ are respectively the free-surface geopotential and horizontal velocity of a shallow fluid covering a horizontal domain S; $\mathbb{n}$ is a vertical unit vector, f the Coriolis coefficient, and ∇ the horizontal differential operator. Linearised about the equilibrium state defined by a uniform geopotential ϕ_o and a zero velocity.

$$\phi = \Phi_o \qquad (12a)$$

$$\mathbb{V} = o \qquad (12b)$$

equations (11) become

$$\frac{\partial \delta \phi}{\partial t} + \Phi_o \nabla \cdot \delta \mathbb{V} = o \qquad (13a)$$

$$\frac{\partial \delta \mathbb{V}}{\partial t} + \nabla \delta \phi + \mathbf{n} \times f \, \delta \mathbb{V} = o \qquad (13b)$$

where $\delta\phi$ and $\delta\mathbb{V}$ are now "perturbations" from state (12). We are going to study the convergence of an assimilation when the appropriate linearised perturbation equations are Eqs.(13). According to the developments of the previous subection, this is the case in either of the following situations

- the time evolution of the system under observation is given by the non-linear equations (11), the "observed solution" being the state of rest (12).

- the time evolution of the system under observation is given by the linear equations (13), the "observed solution" being any solution of these equations.

Despite their simplicity, considering these cases leads to a number of instructive conclusions which, as will be seen below, remain valid in more complex and more realistic situations.

It turns out not to be necessary to determine explicitly the matrix R_y^y corresponding to equations (13) in order to conclude as to the convergence of an assimilation. For any solution of equations (13), the following quadratic quantity

$$E = \frac{1}{2} \int_S (\phi^2 + \Phi_o \mathbb{V}^2) \, dS \qquad (14)$$

which represents the total energy of the perturbation, is conserved in time (from now on, we will drop the prefixes δ in (13)). The geopotential ϕ_o being necessarily positive in any physically meaningful situation, E is $\geqslant 0$ and can be 0 only if the perturbations ϕ and $\mathbb{V}$ are 0 everywhere on S. Let us consider an assimilation performed on observations of geopotential and/or velocity. In the simplest updating procedure the perturbation ϕ ($\mathbb{V}$) will be set equal to 0 over that part of S over which the geopotential (velocity) has been observed, and will be kept unchanged elsewhere. E cannot consequently increase at the

time of an introduction of observations, and will tend to some limit
$E_\infty \geq 0$ as the assimilation is infinitely continued. If $E_\infty = 0$, the assimilation
will be successful in the sense that it will reconstitute the complete mass and
wind fields of the observed solution. The following theorem is proved in T77.

- In the case of a forward-backward assimilation, or in the case
  of a purely forward assimilation performed on observations which
  have an exactly periodic time distribution, the limit E_∞ will be 0
  for any initial perturbations ϕ and ψ if and only if the available
  observations are numerous enough to define uniquely the observed
  solution.

It is obvious that a necessary condition for convergence of an assimilation is
that only one solution of the basic equations be compatible with the available
observations. The above theorem states that in the case of equations (13), and
because of energy conservation (14), this condition is also sufficient.

Energy conservation thus ensures the convergence of an assimilation performed
with equations (13) but it tells us nothing as to the rapidity of that convergence.
The latter will depend to a large extent on the particular space-time distribution
of the observations. We will consider the case of successive observations of the
complete mass field, separated by a constant time interval.

Assuming the Coriolis coefficient to be constant, and taking the Fourier transform
of Eqs. (13), one obtains for wave vector $\mathbf{k}$.

$$\frac{d\phi}{dt} + \Phi_0 D = 0 \tag{15a}$$

$$\frac{dD}{dt} - k^2\phi - f\zeta = 0 \tag{15b}$$

$$\frac{d\zeta}{dt} + f D = 0 \tag{15c}$$

where D and ζ are respectively the divergence and vorticity of the wind field.
The resolvent matrix of Eqs. (15) between times t and $t + \Delta\tau$, which depends
only on $\Delta\tau$, reads

$$R(\Delta\tau) = \begin{pmatrix} \gamma^2+(1-\gamma^2)\cos\beta & -\dfrac{\Phi_o}{\alpha}\sin\beta & -\dfrac{\Phi_o}{\alpha}\gamma(1-\cos\beta) \\ \dfrac{lk^2}{\alpha}\sin\beta & \cos\beta & \gamma\sin\beta \\ -\dfrac{lk^2}{\alpha}\gamma(1-\cos\beta) & -\gamma\sin\beta & 1-\gamma^2(1-\cos\beta) \end{pmatrix} \quad (16)$$

where α is the frequency of inertia-gravity waves

$$\alpha = \sqrt{f^2 + lk^2 \Phi_o} \quad (17)$$

and β and γ are defined by

$$\beta = \alpha\Delta\tau \quad (18a)$$

$$\gamma = \dfrac{f}{\alpha} \quad (18b)$$

The parameter γ, which is always comprised between 0 and 1, is a measure of the ratio of the scale defined by lk to the Rossby radius of deformation $Ro = \dfrac{\sqrt{\Phi_o}}{f}$; γ is close to 1 for scales large compared to Ro and close to 0 for scales small compared to Ro.

We will assume that the complete mass field has been observed at successive times separated by $\Delta\tau$, and that the observations are introduced into the numerical integration without modification of the wind field. The mass field ϕ and the wind field $\binom{D}{\zeta}$ therefore stand respectively for the vectors X and Y of Section 1. An important remark can be made at this point: since the geopotential appears in the divergence equation (15b), but not in the vorticity equation (15c), an introduction of mass observations performed without modifying the wind field will modify the first time derivative of D, but only the second time derivative of ζ. We can therefore expect an assimilation of that type to generally have a stronger effect on the divergent component of the wind field than on its rotational component.

From formula (16) the expression for the matrix $R_Y^Y(\Delta\tau)$ is

$$R_Y^Y(\Delta\tau) = \begin{pmatrix} \cos\beta & \gamma\sin\beta \\ -\gamma\sin\beta & 1-\gamma^2(1-\cos\beta) \end{pmatrix} \quad (19)$$

We will successively consider the cases of a purely forward assimilation and of a forward-backward assimilation. Unless it is necessary to distinguish between different values of the argument, we will normally drop the argument $\Delta\tau$.

2.2.1 Forward assimilation

The wind difference is multiplied at each assimilation step by the matrix R_y^y. The rate of convergence will therefore depend on the spectral radius $\rho(R_y^y)$. The variations of $\rho(R_y^y)$ with the parameters β and γ are shown on Fig. 1. The spectral radius is everywhere ≤ 1, as could a priori be deduced from the energy conservation (14). It is equal to 1 for the values $\beta = \ell\pi$ (ℓ integer), $\gamma = 0$, $\gamma = 1$ and for those values only which, in agreement with the theorem stated above, correspond to cases when the available observations do not uniquely define the observed solution.

For $\beta = \ell\pi$, a "stroboscopic" effect between the intrinsic period $\frac{2\pi}{\alpha}$ of equations (15) and the observing period $\Delta\tau$ makes the successive observations redundant.

For $\gamma = 1$, R_Y^Y is a rotation matrix of angle β (Eqn. (19)) whose both eigenvalues have modulus 1. Neither the divergence nor the vorticity is then reconstructed by the assimilation. For values of γ close to 1, both the divergence and the vorticity are reconstructed at a slow rate (see T77).

In a meteorologically realistic barotropic model, the constant Φ_o must be about $9 \times 10^4 \, m^2 s^{-2}$. The value of γ is then approximately .3 for the largest scales, and tends to 0 as the scale decreases. It is therefore for values of γ smaller than .3 that a detailed study of the assimilation is most appropriate.

When $\gamma = 0$ (i.e. when $f = 0$), the rotational component of the wind is disconnected from the rest of the flow (Eqs. (15)). An assimilation will then reconstruct only the divergent component of the wind, as can be seen from expression (19). For small, non-zero values of γ, both the divergent and the rotational components will be reconstructed, but the former will be reconstructed more rapidly. After an introduction of mass observations, the proportion of the energy of the difference flow contained in the geostrophic mode is equal to

$$E_g = (1-\gamma^2) \frac{|\zeta|^2}{|\zeta|^2 + |D|^2}$$

It is at most equal to $1 - \gamma^2$. As the number of data introductions increases to infinity, and the difference between the model and the observed solutions tend to 0, E_g will either tend to a limit or oscillate between 0 and $1-\gamma^2$, depending on the values of β and γ. Fig. 2 shows that for small values of γ, E_g will tend to a limit which is close to 1. This means that the gravity wave component of the flow will be reconstructed much more rapidly than the geostrophic component. This is in complete contradiction with an often made hypothesis according to which the main difficulty in data assimilation is to get rid of the spurious gravity waves excited by the introduction of observations. Indeed, what the assimilation will do in the present case is to reconstruct rapidly the

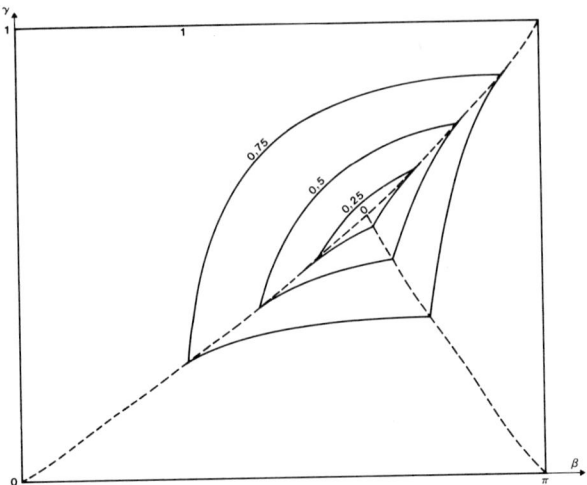

Fig.1 Variations of the spectral radius of the matrix R_y^y with respect to the parameters β and γ. Because of symmetries, only the range $0 \leqslant \beta \leqslant \pi$ is considered.

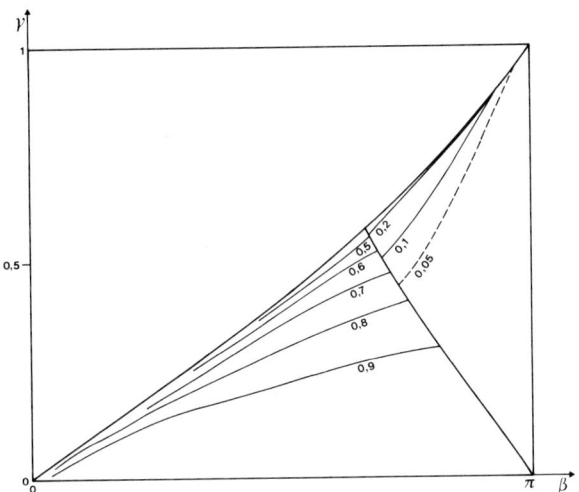

Fig. 2 Variations of the asymptotic value of the ratio Eg with respect to the parameters β and γ. In the blank area, which corresponds to $\gamma > \tan\frac{\beta}{4}$, Eg has no limit and oscillates between 0 and $1-\gamma^2$.

gravity wave component of the flow (possibly 0, if the observed solution is geostrophic) and slowly the geostrophic component.

Figs. 3a and 3b show the spectra of the divergence and vorticity differences respectively at the beginning and after a given time of forward assimilation. The assimilation was performed with a non-linear spherical barotropic model (described in T77), the "observed" solution being the state of rest (12), with $\Phi_o = 8.13 \times 10^4 \, m^2 s^{-2}$. The features anticipated above are clearly apparent. Both the divergence and the vorticity are reconstructed at all wavelengths. The former is reconstructed more rapidly, especially at small wavelengths (small γ's). Wavelengths such that $\beta = \ell\pi$, ℓ integer, have been indicated. The decrease of the difference is slower for these wavelengths.

The effect of latitude on the assimilation is visible from Fig. 4, which shows the time variations of the divergence and vorticity differences, at both the equator and high latitudes. The divergence difference decreases rapidly, at a rate independent of latitude, while the vorticity difference decreases slowly at high latitude, and does not decrease at all at the equator, where the value of γ is 0.

2.2.2 Forward-backward assimilation

In one cycle of forward-backward assimilation performed on N successive observations of the mass field separated by $\Delta\tau$, the wind difference $\binom{D}{\zeta}$ is multiplied N-1 times by $R_y^y(-\Delta\tau)$ in the backward phase of the cycle, and N-1 times by $R_y^y(\Delta\tau)$ in the forward phase. The amplification matrix A of (10) is therefore

$$A = [R_y^y(\Delta\tau)]^{N-1} [R_y^y(-\Delta\tau)]^{N-1}$$

From the energy conservation (14), we can tell that necessarily $\rho(A) \leq 1$. For $N > 2$ (if N=2, $\rho(A) = 1$ for all values of β and γ because the uniqueness condition stated above is not satisfied), the variations of $\rho(A)$ with β and γ are similar to those shown on Fig. 1. In particular $\rho(A) = 1$ for the same limiting values. The conclusions already drawn remain valid.

An interesting question is the following: for a given time distribution of observations, will the convergence be more rapid in a forward-backward assimilation, or in a purely forward assimilation? In one cycle of forward-backward assimilation, the wind difference is multiplied by the above matrix A. In the same time, i.e. $2(N-1)\Delta\tau$, of forward assimilation, the wind difference is multiplied by $C = [R_y^y(\Delta\tau)]^{2(N-1)}$. The question considered is therefore equivalent to comparing the spectral radii $\rho(A)$ and $\rho(C)$. Eqs. (18a) and (19) show

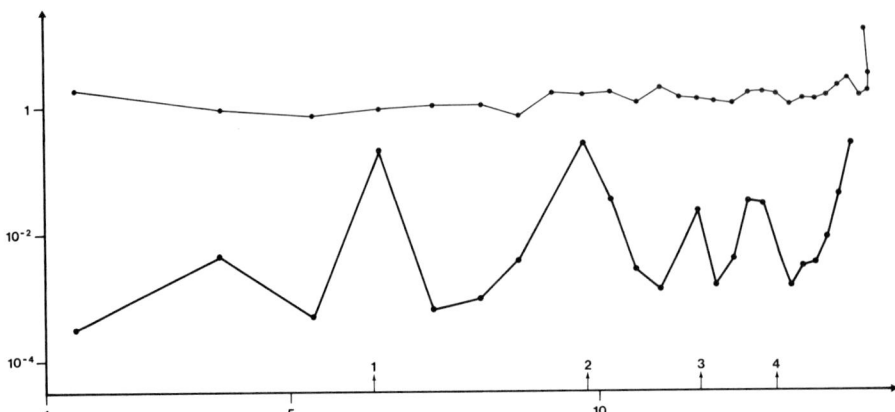

Fig. 3a Spectra of the divergence difference at the beginning (upper curve) and after 9 days of assimilation (lower curve). The unit in the vertical axis is arbitrary, but the initial difference has been generated by a random perturbation with standard deviation .58 ms^{-1} on each component. The abscissa variable is the logarithm of the modulus of the Laplacian eigenvalue. It mostly depends on the latitudinal wavenumber of the corresponding eigenfunction. The arrows indicate abscissa values such that $\beta = \ell\pi$, ℓ integer, together with the corresponding values of ℓ.

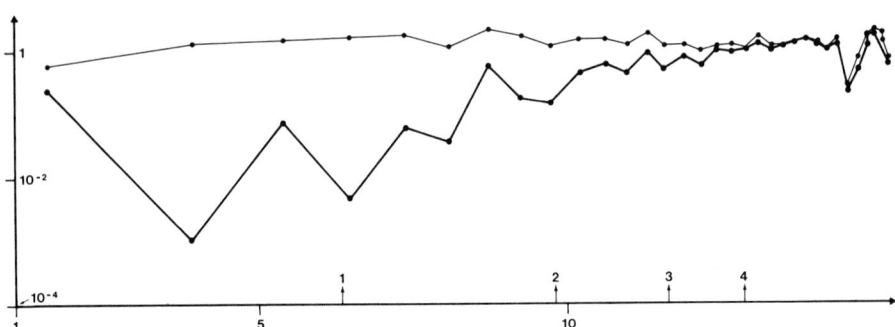

Fig.3b Same as 3a, for the vorticity difference

that $R_y^y(-\Delta\tau)$ is the transpose of $R_y^y(\Delta\tau)$, so that A and C are respectively equal to $B\tilde{B}$ and B^2, where B is the matrix $[R_y^y(\Delta\tau)]^{N-1}$ and $\tilde{B}$ is the transpose of B. Now a result of matrix algebra (see. e.g. Varga (1962)) tells us that, for any matrix B, $\rho(B\tilde{B})$ is the square of the Euclidean norm of B. Consequently

$$\rho(B^2) \leq \rho(B\tilde{B}) \tag{20}$$

which, in the present case, means that a purely forward assimilation will converge at least as rapidly as a forward-backward assimilation.

Fig. 5 shows the time variations of the rms wind differences in assimilations of both types performed on data with the same time distribution. In the beginning, the decrease is more rapid for the purely forward assimilation, in agreement with inequality (20). Later on, however, the difference starts increasing in the forward assimilation. This increase, which is in contradiction with the energy conservation (14) can be traced to the fact that energy is conserved by the model only to first order with respect to the time discretisation increment.

2.3 Generalisation to the linearised primitive equation

The foregoing results can be extended to the linearised multi-level primitive equations. These equations conserve with time a positive definite quadratic energy, which is a function of surface pressure, potential temperature and wind (see T77). This energy plays exactly the same role as (14) and it results that an assimilation performed on observations of surface pressure, potential temperature and/or wind will converge under the only condition that the available observations uniquely define the corresponding solution.

In the case of observations of the complete mass field (or wind field), a quantitative study can be made by separating the linearised equations into their vertical modes, the time evolution of each of which is given by Eqs. (15), Φ_o being replaced by the appropriate equivalent geopotential. The equivalent geopotential decreases, and the parameter γ consequently increases (Eqs. (17) and (18b)) as the order of the vertical mode increases. According to the analysis of the previous subsection, we can therefore expect that the difference between the rate of reconstitution of divergence and vorticity will be less marked for modes of higher order. Figs. 6a and 6b, which are in the same format as Figs. 3a and 3b refer to the internal mode of a two-level model, for which the equivalent geopotential was $\Phi_1 = 1400 \text{ m}^2\text{s}^{-2}$. At large scales, divergence and vorticity are reconstituted at about the same rate. At small scales divergence is still reconstituted more rapidly than vorticity, but the latter is reconstituted more rapidly than on the external mode (see Fig. 3a).

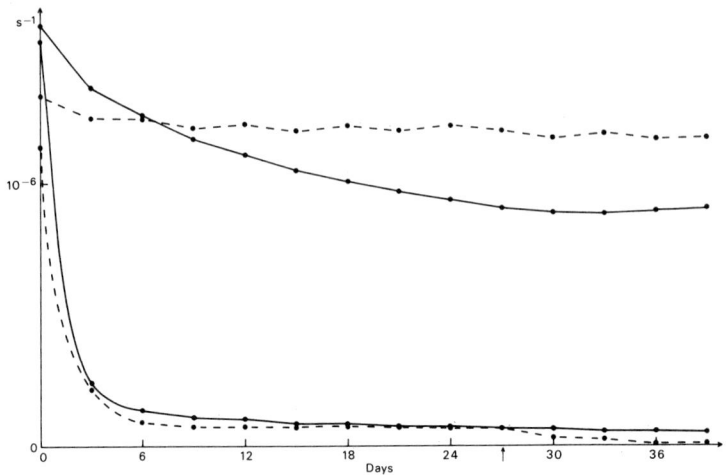

Fig.4 Variations with time of the rms differences of
- high latitude divergence (lower solid curve)
- equatorial divergence (lower dotted curve)
- high latitude vorticity (upper solid curve)
- equatorial vorticity (upper dotted curve)
The scale on the vertical axis is linear

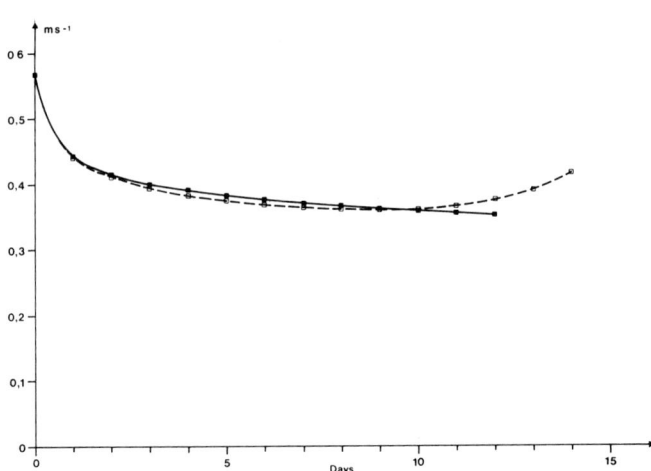

Fig.5 Variations of the rms wind difference in a forward-backward assimilation (solid curve) and a purely forward assimilation (dotted curve), both with the same space-time distribution of observations.

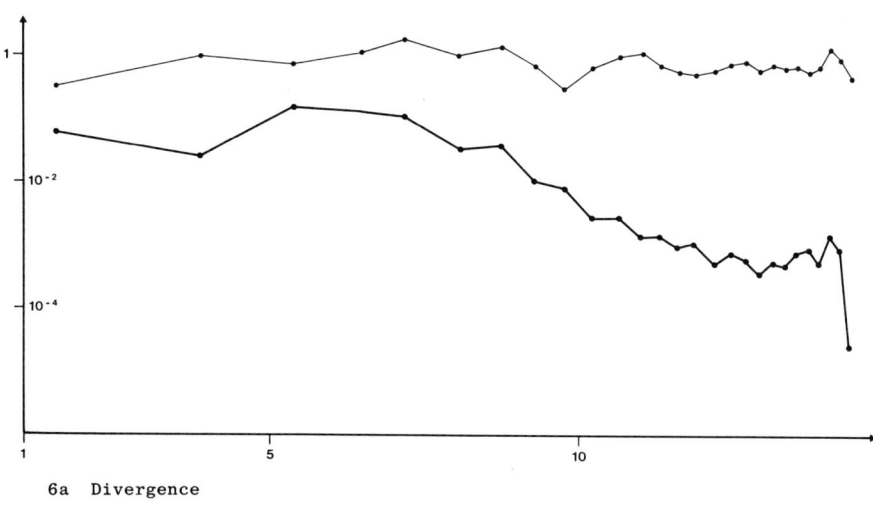

6a Divergence

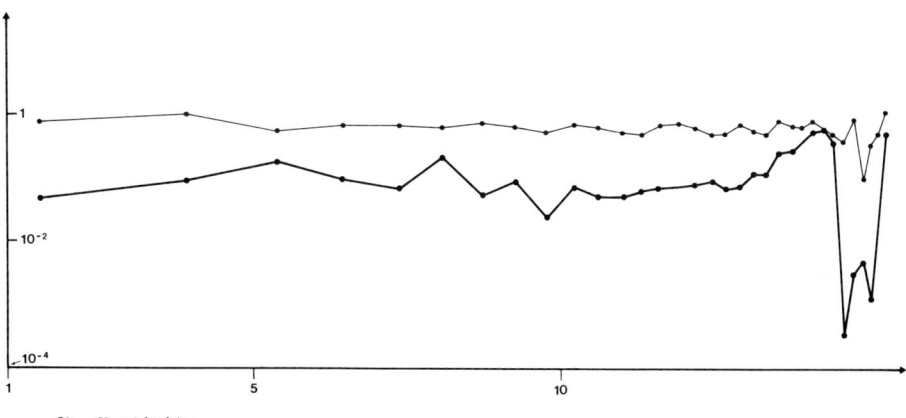

6b Vorticity

Fig. 6 Same as Fig. 3 for the internal mode of a two-level model

2.4 The effect of geostrophic adjustment

The results established so far are valid for any solution of equations (15), whether this solution is geostrophic or not. It has often been hypothesised that geostrophic adjustment, whose effect is to reestablish geostrophic balance after it has been disrupted by an introduction of observations, is a necessary ingredient for the success of an assimilation procedure. Our results show that such is not the case. We will come back later to this point, and will here study how the results established so far are modified by the presence of a geostrophic adjustment process. We will assume this process (whose exact mechanism is irrelevant in the present context) to be such that after each introduction of observations, and before the next introduction, the gravity wave component of the flow is completely damped, while the geostrophic component is not altered in any way. This simple "model" of geostrophic adjustment, has already been used by several authors (e.g. by Williamson and Dickinson (1972)).

The energy E of the difference flow will now decrease both when observations are introduced, as before, and between introductions because of the damping of gravity waves. However, since the energy decrease can be distributed between the various component of the flow, this decrease will not necessarily be more rapid than in the absence of geostrophic adjustment.

Acting on the difference $(\phi = 0, D, \zeta)$ resulting from an introduction of observations, geostrophic adjustment will leave only the corresponding geostrophic components, viz.

$$\phi_g = -\frac{\Phi_o}{\alpha} \gamma \zeta$$

$$D_g = 0$$

$$\zeta_g = (1-\gamma^2)\zeta$$

This component is stationary with time, and the next introduction of observations will result in ϕ_g being set equal to 0. The wind difference $\binom{D}{\zeta}$ will therefore be multiplied at each assimilation step by the following matrix

$$G = \begin{pmatrix} 0 & 0 \\ 0 & 1-\gamma^2 \end{pmatrix} \qquad (21)$$

The spectral radius of G, equal to $1-\gamma^2$, is always less than 1, except for $\gamma=0$. Fig. 7 shows the variation of the difference $\rho(R_y^y) - \rho(G)$ with β and γ.

- for large values of γ, this difference is positive, which means that an assimilation will converge more rapidly if geostrophic adjustment is present. This can be easily explained: for these values of γ (i.e. for scales large compared with the Rossby radius of deformation) the effect of geostrophic adjustment is to adjust the wind field to the mass field, the latter not being modified. Geostrophic adjustment can therefore do what simple assimilation cannot do in this particular case, namely reconstruct a wind field consistent with the observed mass field.

- for small values of γ on the contrary (which are relevant to atmospheric barotropic models), and except in the vicinity of the values $\beta = 0$ and $\beta = \pi$, an assimilation is more efficient if no geostrophic adjustment is present. This too can easily be explained: for small values of γ, geostrophic adjustment results in the mass field being adjusted to the wind field, so that mass observations are in effect rejected. For these values, an assimilation without geostrophic adjustment converges slowly, but geostrophic adjustment only makes the convergence still slower.

3. COMPARISON WITH RESULTS OBTAINED WITH THE NON-LINEAR EQUATIONS

It is interesting to compare the theoretical results obtained above with numerical results obtained using a non-linear model. Figs. 8 and 9, which are in the same format as Figs. 3 and 6, are relative to an assimilation performed with non-linear equations. The model used, and the space time distribution of observations, were the same as before, but the "observed" solution was a fully non-linear meteorologically realistic solution. In addition, the model contained a divergence dissipation, acting as geostrophic adjustment (see Sadourny (1972)).

Figs. 8 and 9 refer respectively to the external and internal modes of the model. The main qualitative features deduced from Eqs. (15) are still present. Both the divergence and the vorticity are reconstructed at all wavelengths and for both vertical modes. The divergence is reconstructed more rapidly, especially at small wavelengths and for the external mode. All stroboscopic effects have disappeared, which corresponds to the fact that non-linearity destroys any periodicity in the flow. There is moreover an important change from the linear case: the global rate of reconstitution of the wind field is now much more rapid (Fig. 10). This fact may be due in part to the dissipation

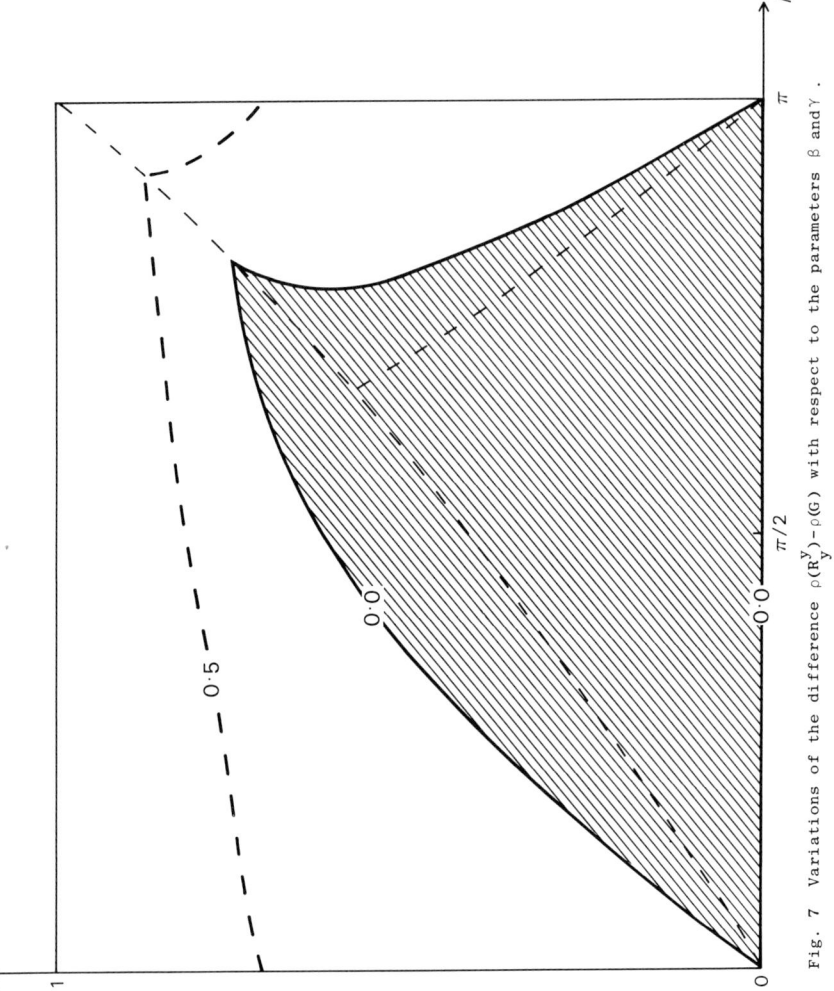

Fig. 7 Variations of the difference $\rho(R_y^v) - \rho(G)$ with respect to the parameters β and γ. The difference is negative in the shaded area, positive elsewhere.

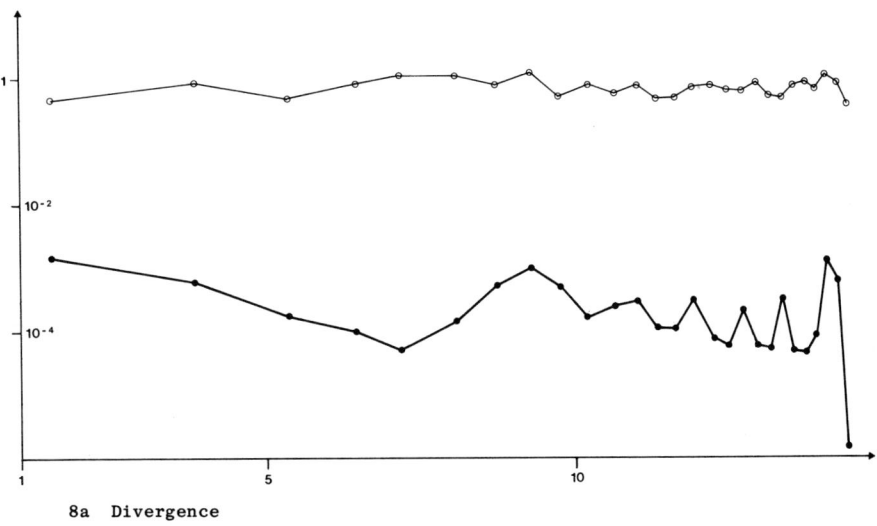

8a Divergence

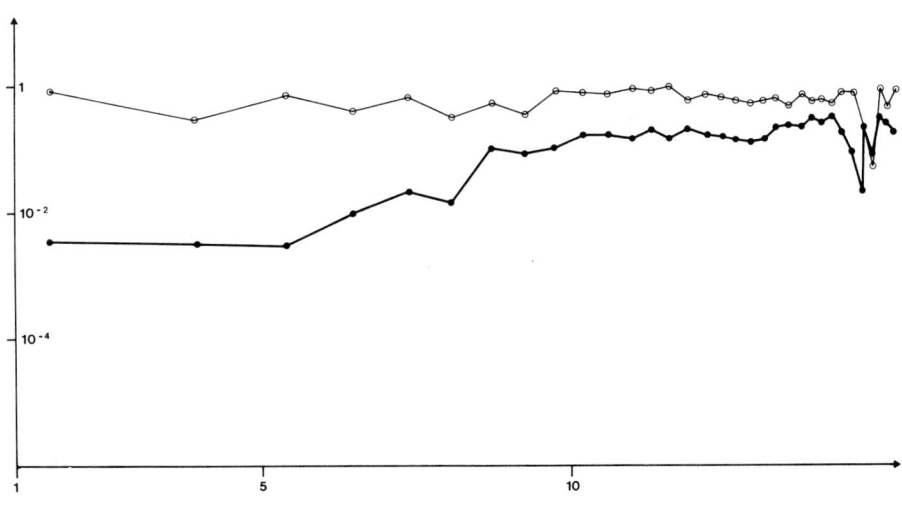

8b Vorticity

Fig. 8 Same as Fig. 3 for the external mode of a fully non-linear solution

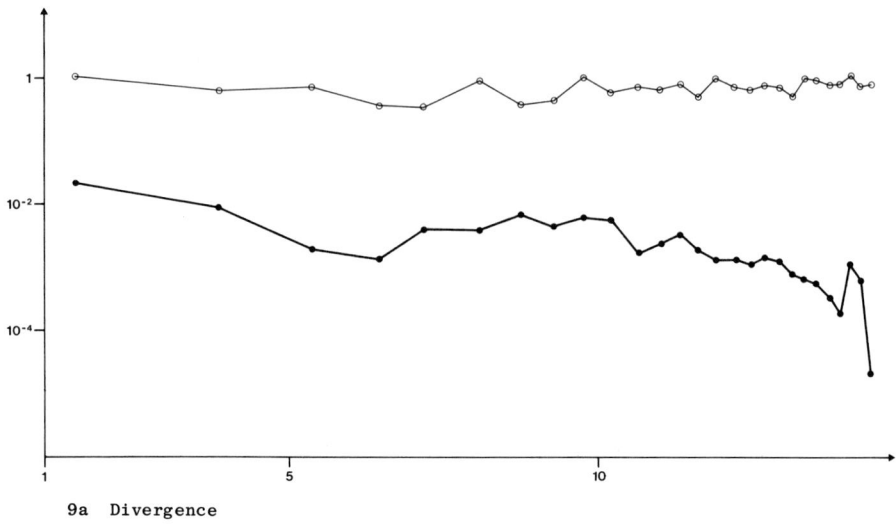

9a Divergence

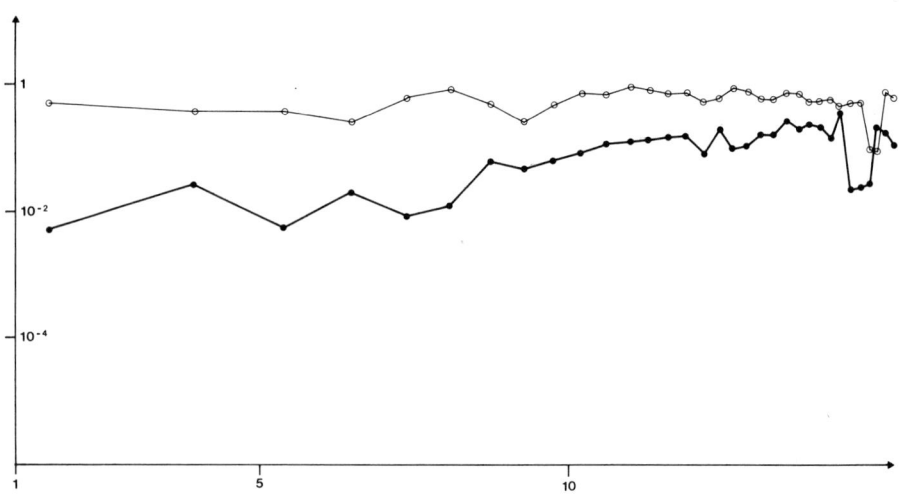

9b Vorticity

Fig. 9 Same as Fig. 3 for the internal mode of a fully non-linear solution

of gravity waves, but cannot be due only to that dissipation, since it remains true for small values of the parameter γ, for which, as seen in 2.4 dissipation of gravity waves will not speed up the rate of convergence. For instance, the equatorial vorticity is now reconstructed (Fig.11) while it was not in the linear case (Fig. 4). Non-linear advection must therefore be at least in part responsible for the speeding up of convergence.

4. THE CASE OF SUCCESSIVE OBSERVATIONS CLOSE IN TIME

The explicit determination of the amplification matrix (10) and/or of its spectral radius will most often be difficult, if not impossible, since it will normally require the explicit knowledge of the linearised perturbation Eqs. (3) in the vicinity of the observed solution. In the case of the meteorological equations, it is unlikely that Eqs. (3) can be determined, except for a few explicitly known solutions. However, when the successive observations are close in time, the amplification matrix and the corresponding convergence criterion assume simple forms, depending only on conditions local in time. We will now proceed to the study of this simpler case.

4.1 The simplified convergence criterion

We will first assume that two observations of X are available, at times t_1 and t_2. The amplification matrix (10) is

$$A = R_y^y(t_2,t_1) \, R_y^y(t_1,t_2)$$

where we do not use the auxiliary time τ any more. The reversibility equation (8) together with decomposition (5) imply

$$R_y^y(t_2,t_1) \, R_y^y(t_1,t_2) + R_y^x(t_2,t_1) \, R_x^y(t_1,t_2) = I$$

where I is now the unit matrix of order q.

Therefore

$$A = I - R_y^x(t_2,t_1) \, R_x^y(t_1,t_2) \tag{22}$$

Equations (3) imply in turn

$$R_y^x(t+\Delta t,t) = \Delta t \, \frac{DG}{DX}(t) + o(\Delta t)$$

$$R_x^y(t+\Delta t,t) = \Delta t \, \frac{DF}{DY}(t) + o(\Delta t)$$

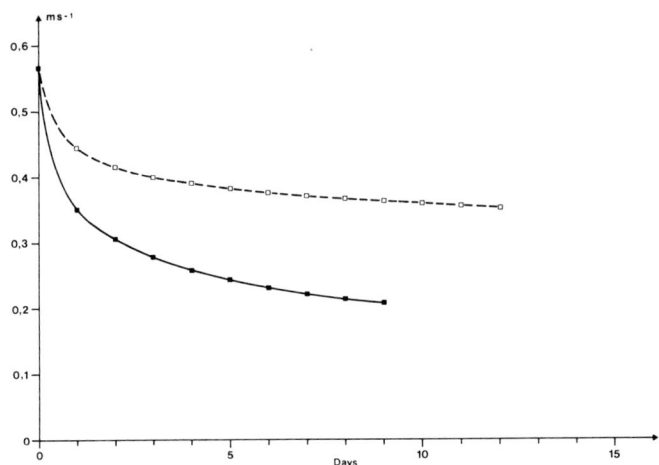

Fig. 10 Variations with time of the global wind rms difference in the non-linear (solid curve) and linear (dotted curve) cases respectively.

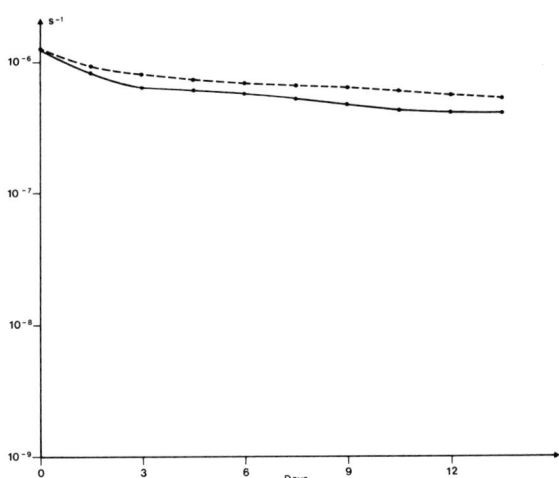

Fig. 11 Variations of the rms equatorial vorticity difference in the non-linear case. Solid curve: external mode. Dotted curve: internal mode.

Setting $\Delta t = t_2 - t_1$, and carrying these expressions into (22) lead to

$$A = I + \Delta t^2 \frac{DG}{DX}(t_1) \frac{DF}{DY}(t_2) + o(\Delta t^2)$$

To order Δt, the matrices $\frac{DG}{DX}$ and $\frac{DF}{DY}$ can be taken indifferently at time t_1 or t_2.

We shall write

$$A = I + \Delta t^2 \frac{DG}{DX} \frac{DF}{DY} + o(\Delta t^2) \tag{23}$$

with no more precision.

Let λ_j (j=1,...,q) be the eigenvalues of the qxq matrix $\frac{DG}{DX}\frac{DF}{DY}$. The eigenvalues of A are

$$a_j = 1 + \lambda_j \Delta t^2 + o(\Delta t^2) \qquad (j=1,\ldots,q)$$

The convergence criterion derived above is that in the complex plane all the eigenvalues a_j lie inside the circle (C) centered at the origin with radius unity. This condition is verified if all the eigenvalues λ_j have a strictly negative real part (Fig. 12).

$$\mathcal{R}(\lambda_j) < 0 \qquad (j=1,\ldots,q) \tag{24}$$

The matrix $\frac{DG}{DX}\frac{DF}{DY}$ is the product of two matrixes, each of which represents the dependence of the time evolution of one of the two parts X or Y with respect to the other. The matrix $\frac{DG}{DX}\frac{DF}{DY}$ therefore represents in some sense the coupling between the time evolutions of X and Y. It will be called the coupling matrix between X and Y.

In the case when X has been observed at N + 1 successive times separated by a constant time interval Δt, it can be shown (e.g. by induction on N) that the amplification matrix is

$$A = I + N \Delta t^2 \frac{DG}{DX}\frac{DF}{DY} + o(\Delta t^2) \tag{25}$$

which leads to the same convergence criterion (24).

4.2 Application to the shallow-water equations

Let us consider the shallow-water equations (11), the geopotential ϕ and the velocity $\mathbb{V}$ standing respectively for the vectors X and Y, and equations (11a) and (11b) for equations (1a) and (1b) respectively. For the previous developments to be valid, we must assume equations (11) to have been discretised to a finite number of parameters. However, for the sake of simplicity, we will keep notations usual for a continuum.

The velocity field $\mathbb{V}$ appearing linearly in Eq.(11a), the equivalent of the jacobian matrix $\frac{DF}{DY}$ is now the linear operator L_1 which to any "perturbation" $\partial \mathbb{V}$ of the wind field, associates the scalar field defined by

$$L_1(\partial \mathbb{V}) = -\nabla \cdot (\phi \, \partial \mathbb{V})$$

Similarly, the geopotential ϕ appears linearly in Eq. (11b) and the equivalent of the jacobian matrix $\frac{DG}{DX}$ is the linear operator L_2 which, to any perturbation $\partial \phi$ of the geopotential field, associates the vector field defined by

$$L_2(\partial \phi) = -\nabla(\partial \phi)$$

The equivalent of the coupling matrix $\frac{DG}{DX}\frac{DF}{DY}$ is the linear operator obtained by composing L_1 with L_2, i.e. the operator C which, to any wind field perturbation $\partial \mathbb{V}$ associates the following vector field

$$C(\partial \mathbb{V}) = L_2[L_1(\partial \mathbb{V})] = \nabla[\nabla \cdot (\phi \partial \mathbb{V})] \qquad (26)$$

The next step is to determine the signs of the real parts of the eigenvalues of C. First we introduce the scalar product defined for any two wind field perturbations $\partial \mathbb{V}$ and $\partial \mathbb{V}'$ (with possibly complex components) by

$$\langle \partial \mathbb{V}, \partial \mathbb{V}' \rangle = \int_S \phi \; \partial \mathbb{V} \cdot \partial \mathbb{V}'^{*} \, dS \qquad (27)$$

where $\partial \mathbb{V}'^{*}$ is the complex conjugate of $\partial \mathbb{V}'$. Since the geopotential is positive everywhere on the domain S, (27) defines a scalar product. Then, for any two two perturbations $\partial \mathbb{V}$ and $\partial \mathbb{V}'$

$$\langle \partial \mathbb{V}, C(\partial \mathbb{V}') \rangle = \int_S \phi \; \partial \mathbb{V} \cdot \nabla[\nabla \cdot (\phi \partial \mathbb{V}^{*})] \, dS = -\int_S \nabla \cdot (\phi \partial \mathbb{V}) \; \nabla \cdot (\phi \partial \mathbb{V}'^{*}) \, dS \qquad (28)$$

The latter expression is symmetrical with respect to $\delta \mathbb{V}$ and $\delta \mathbb{V}'^{*}$, which means that the operator C is self-adjoint with respect to the scalar product (27).

Its eigenvalues are consequently real. Let λ be one of these eigenvalues. Setting $\delta \mathbb{V}' = \delta \mathbb{V}$ in (28), where $\delta \mathbb{V}$ is an eigenfunction associated with λ, leads to

$$\lambda < \partial \mathbb{V}, \partial \mathbb{V}> = - \int_S [\nabla \cdot (\phi \, \partial \mathbb{V})]^2 \, dS$$

which shows that λ is negative except if

$$\nabla \cdot (\phi \, \partial \mathbb{V}) = 0 \quad \text{everywhere on } S \tag{29}$$

It is obvious from (28) that any perturbation $\partial \mathbb{V}$ which satisfies this condition is an eigenfunction of C, associated with eigenvalue 0. All the eigenvalues of C are therefore negative, except one which is 0.

If we ignore for the time being the $o(\Delta t^2)$ term in (23), it results than an assimilation performed on successive observations of the geopotential separated by Δt will reconstitute the windfield except for a residual error verifying (29). More precisely, taking the vorticity of (26) shows that the vorticity of the wind difference is not modified in an assimilation cycle. The residual error $\delta \mathbb{V}_\infty$ will therefore be defined by the following two conditions: it satisfies (29) and its vorticity is equal to the vorticity of the wind difference at the beginning of the assimilation.

These results are true only if Δt is small enough so that no eigenvalue of the amplification matrix (23) has modulus larger than 1. Still ignoring the $o(\Delta t^2)$ term, this condition reads

$$- 1 < 1 + \lambda_M \, \Delta t^2$$

or

$$\Delta t < \sqrt{- \frac{2}{\lambda_M}} \tag{30}$$

where λ_M is the (negative) eigenvalue of C with the largest modulus. It is not difficult to see that condition (30) is basically the same as the Courant-Friedrichs-Lewy condition for stability of a numerical integration of Eqs. (11). It leads for Δt to values of the same magnitude, typically $\Delta t < 15$ mn for ordinary spatial resolutions.

The additional term $o(\Delta t^2)$ in (23) turns out to be too complex to be studied analytically, and numerical experiments have been performed in order to determine if, and how, it modifies the above results. Since the total number of parameters of a shallow-water equation model is three times the number of parameters defining the complete mass field, two successive observations of the latter cannot in any case define the complete state of the flow. Accordingly, the experiments were performed with three successive observations of ϕ, separated by Δt. Development (23) is then replaced by development (25) with N=2. Condition (30) is replaced by the still stricter condition

$$\Delta t < \sqrt{\frac{1}{\lambda_M}} \equiv \Delta t_c \qquad (31)$$

For the particular model used in the present experiments, the value of Δt_c was about 12.6 mn. The time discretisation increment used was

$$\Delta \tau = 12 \text{ mn} \simeq .95 \, \Delta t_c$$

As for the time interval Δt between successive observations of ϕ, two values were used, one

$$\Delta t = \Delta \tau$$

which satisfies condition (31), and the other

$$\Delta t = 3 \, \Delta \tau \simeq 2.85 \, \Delta t_c$$

which does not.

Fig. 13 shows the variations of the root-mean-square wind difference in the two experiments. The difference decreases in both cases and more rapidly, for the same number of assimilation cycles, with the larger value of Δt. This means that, when Δt reaches the limit value Δt_c, the $o(\Delta t^2)$ term is no more negligible, and indeed is such as to decrease further the spectral radius of the amplification matrix. An additional fact of interest can be seen from Fig. 14, which shows the variations of the rms vorticity difference in the same two assimilations. As said above, the vorticity is not modified by the first two terms of development (25), and only the $o(\Delta t^2)$ term can account for a possible variation of the vorticity difference. Fig. 14 shows that this difference decreases for both values of Δt, and more rapidly for the larger value.

It thus appears that, at least in the case of the numerical model used here, the $o(\Delta t^2)$ term contributes to the convergence of an assimilation and, particularly, leads to the reconstitution of the rotational part of the wind field. The basic mathematical reason for this has not been rigorously

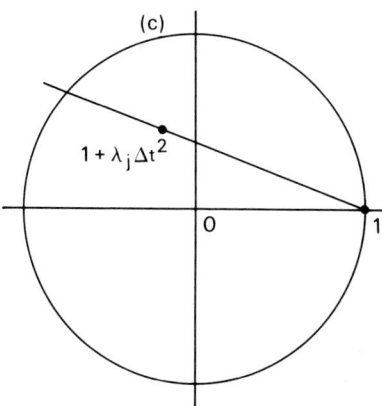

Fig. 12 Graphical illustration of condition (24). For small Δt, the point $1 + \lambda_j \Delta t^2 + o(\Delta t^2)$ lies inside circle (C) if the real part $\mathfrak{R}(\lambda_j)$ of λ_j is strictly negative, and outside (C) is $\mathfrak{R}(\lambda_j)$ is strictly positive. When $\mathfrak{R}(\lambda_j) = 0$, the point may lie inside, on or outside circle (C), depending on $o(\Delta t^2)$

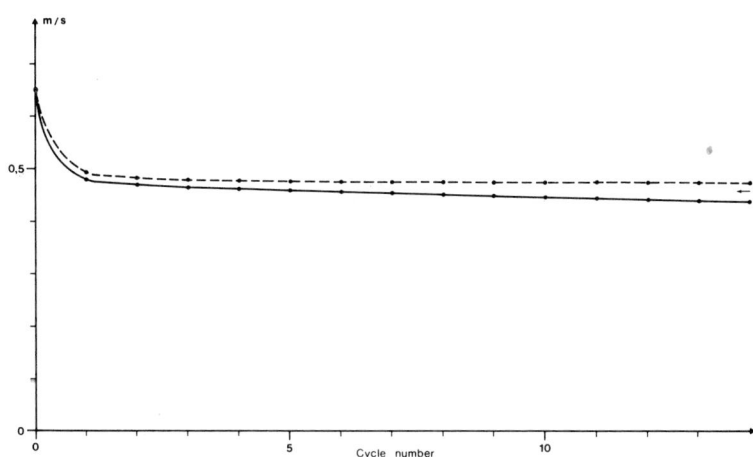

Fig. 13 Variations, as functions of the number of assimilation cycles, of the rms wind difference in two assimilations of mass observations. The upper curve corresponds to a time interval between observations $\Delta t = \Delta t_c$, and the lower curve to $\Delta t = 3\Delta t_c$.

established, but two facts strongly suggest that the role of the Coriolis
acceleration is here fundamental. First, it has been shown in Section 2 that,
in the case of the linearised equations, it is the Coriolis acceleration,
which is then the only interaction between divergence and vorticity, which
ensures the reconstitution of vorticity. The second fact is apparent from
Fig. 15, which shows, as functions of latitude, the proportion of the initial
rms difference remaining after a given time of assimilation on divergence
and vorticity respectively (this figure refers to the assimilation performed
with the larger value $\Delta t = 3\Delta\tau$). The rate of reduction of the divergence
difference is rapid and exhibits no variation with latitude. The rate of
reduction of the vorticity difference, on the contrary, is slow, particularly
in low latitudes. This suggests that the reconstitution of vorticity depends
to a large extent on the Coriolis parameter. It must be noted however that
the vorticity difference is reduced at all latitudes, including at the equator.
In Sections 2 and 3, the same fact was observed only with non-linear
equations, and was ascribed to an additional effect of advection. The same
explanation probaly applies also here.

It is noteworthy that, as in the case of the linear equations, the
numerical convergence of an assimilation procedure is independent of any
specific feature of the solution to be reconstituted and, in particular, of
whether or not this solution is geostrophic. It does not require either
the presence in the assimilating model of any dissipative process, intended
at restoring the geostrophic balance after it has been disrupted by the
introduction of observations. These facts are obvious for the theoretical
results, which have been derived without any particular hypothesis about the
"observed" solution, and without assuming the presence of any dissipative
process. As for the numerical results just presented, they have been obtained
from an "observed" solution which had been produced by numerical integration
of the inviscid equations (11). For some reason, this integration produced a
rather large amount of gravity waves. This in no way prevented the reconstitution of both the divergent and rotational parts of the wind field.

Because of the particular conditions under which the results presented in this
Section have been obtained (small time interval between successive observations,
no dissipation) they are not directly relevant to the practical problem of
assimilation. But, together with the results obtained for the linear equations,
they provide a clear description of the processes at play in an assimilation
of mass observations. First, the direct influence of data introduction upon
the divergence results in a rapid reconstitution of the latter. The

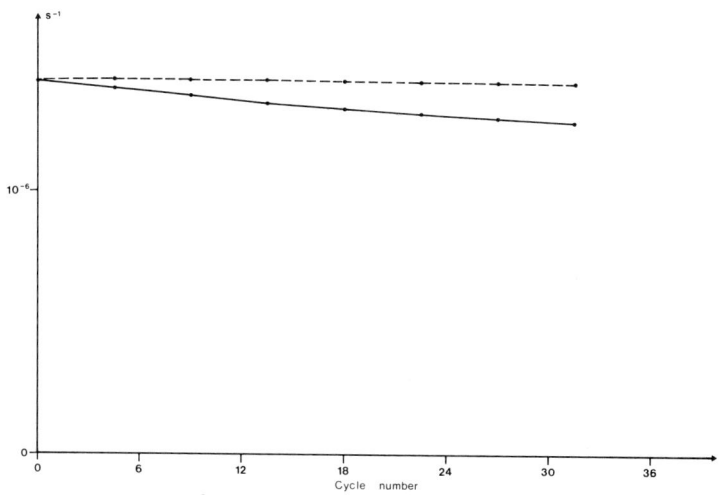

Fig. 14 Same as Fig. 13, for the variations of the rms vorticity difference. The scale on the vertical axis is linear

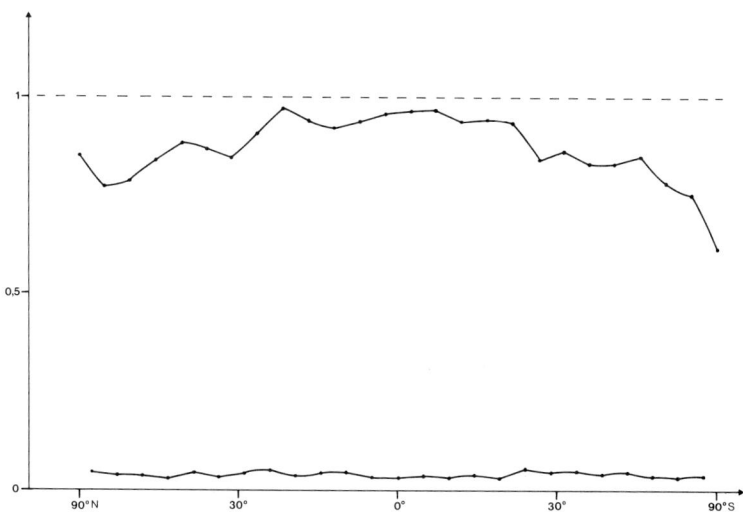

Fig. 15 Latitudinal distribution of the proportion of initial rms difference remaining after a given time of assimilation.
Upper curve: vorticity rms difference.
Lower curve: divergence rms difference

vorticity field is also reconstructed, but indirectly and more slowly, through Coriolis acceleration and, to a lesser extent, through advection.

As for the somewhat unexpected fact that neither geostrophicity nor the presence of a geostrophic adjustment process are basically necessary for the convergence of an assimilation, two conclusions at least can be drawn from it.

- a usual interpretation is that it is because of geostrophic adjustment that assimilation processes converge at all. Our results show that this interpretation is not correct. Other effects than geostrophic adjustment play a role in assimilation, and these effects must be taken into account in the definition of assimilation procedures.

- an assimilation can relatively easily reconstruct the divergence field from the observed time evolution of the mass field. This is clearly apparent, for instance, from Fig. 15. Indeed, the time evolution of the mass field contains information on the divergence inaccessible through direct measurement. For instance, a surface pressure tendency of 1 $mb.hr^{-1}$ corresponds to a vertically averaged divergence of $3 \times 10^{-7} s^{-1}$, well beyond the accuracy of direct measurements. In all present assimilation procedures, the final values of the divergence are determined mostly by the initialization step, which is intended at suppressing unrealistic gravity waves. But it is not known to which accuracy the real values of the divergence are reconstructed by the initialization. It would be of interest to define a method for reconstructing the divergence with all the accuracy allowed by the observations of the mass field. The results presented above suggest that this is possible.

5. THE EFFECT OF OBSERVING AND MODELLING ERRORS

We have exclusively considered so far the "identical twin" case, in which observations are supposed to be exactly compatible with one solution of the assimilating numerical model. This is not so in reality because of observing and modelling errors. We shall now extend the formalism of Section 2 to the case when such observing and/or modelling errors are present.

First, one particular solution of the assimilating model (this solution need not be defined more precisely at this stage) is chosen as a reference solution. Then, for each observed datum, an "error" is defined as the difference between the datum itself and the corresponding value for the reference solution just chosen. The vector made up of all the errors thus defined will be noted E. With the notation of Section 2, the dimension of E is equal to the product Np of the number N of observation times by the dimension p of the observed vector X.

Denoting again by ΔY_n the Y-difference between the model state and the reference solution at the end of the n-th assimilation cycle, a derivation similar to that of Section 2 leads to the following relationship between ΔY_n and ΔY_{n+1}

$$\Delta Y_{n+1} = A \, \Delta Y_n + BE + o(\Delta Y_n, E) \tag{32}$$

where A and B are matrices with respective dimensions $q \times q$ and $q \times Np$. A is the same matrix as in (9) since (32) must reduce to (9) when E=0. The matrix B, like the matrix A, is entirely determined by the resolvent matrix of the linearised perturbation Eqs. (3) in the vicinity of the reference solution.

Now the question is: when the number n of assimilation cycle increases to infinity, and E remains constant, how will ΔY_n behave? The following result is proved in T77: if the spectral radius of A is strictly less than 1, and if E and ΔY_o are small enough, ΔY_n will tend to a limit as n will tend to infinity. However, this limit will not in general be 0.

This result means that, if the observed values remain close enough to one particular model solution, and if the spectral radius of the amplification matrix corresponding to that solution is strictly less than 1, then the assimilation will converge to a limit. However, the corresponding model states at times $t_1, t_2 \ldots t_N$ will not in general lie on one model solution.

The proof given in T77 is too general to provide information of specific interest for the meteorological problem. It does not in particular allow any practical estimate of how "small" E must be for an assimilation to remain convergent. It is probably mostly through numerical experimentation that precise information can be obtained on this point.

6. A THEORETICAL WAY FOR ACCELERATING CONVERGENCE

We assumed in Section 2 that one particular arbitrary choice had been made as for the nature of the vector Y which was kept unchanged when observations of X were introduced. One can wonder how the convergence properties of an assimilation will be modified if it is another vector W, and not Y, which is kept unchanged when X is updated. It is not difficult to show (Talagrand, 1981) that keeping W unchanged is equivalent to adding to Y the following correction whenever X is updated

$$\Delta'Y = D\Delta X + o(\Delta X, \Delta Y) \qquad (33)$$

In this expression D is a q x p matrix, obtained from the dependence of W with respect to X and Y, while ΔX and ΔY are, as before, the X and Y differences between the model and the observed solution. Conversely, any correction of type (33) applied to Y when X is updated keeps unchanged some perfectly defined vector W. We will consider in this section how the convergence properties of an assimilation are modified by a correction of type (33). In fact, only the matrix D will be important for this study and it will not be necessary to consider the corresponding "invariant" vector W.

Many assimilation procedures have been defined, some of which are in operational use, which use a correction of type (33). For a large number of them (but not all) the correction is purely linear, i.e. there is no $o(\Delta X, \Delta Y)$ term. The "optimal analysis" when performed with the model forecast as first guess, as is most often the case (see Lorenc et al.(1977)), is one of them. The matrix D, which is then determined on the basis of statistical considerations is in that case equal to $-QC^{-1}$, where Q is the covariance matrix of ΔY with ΔX, and C the variance-covariance matrix of ΔX with itself. Various schemes, intended at restoring the geostrophic balance disrupted by the introduction of observations, and which are all particular cases of (33), have been defined by Kistler and McPherson (1975) and by Daley and Puri (1980). Several of them have been shown to accelerate the convergence rate of an assimilation. For still another example of (33), see Tadjbakhsh (1969).

The scheme proposed by Kistler and McPherson is of particular interest because its simplicity allows for a ready analysis of its effects. This scheme consists, whenever the mass field is updated, in adding to the wind field a correction which is geostrophically related to the mass correction. The corresponding matrix D is the matrix which expresses the dependence of geostrophic wind on the mass field. Since the global correction thus added to the flow is geostrophic, the gravity wave component of the difference flow is not modified at an introduction time. In a linear model, this component is not modified either between two integrations. The Kistler-McPherson correction used alone in a linear model therefore prevents the convergence of an assimilation, since it maintains the gravity wave component of the difference flow. The situation becomes completely different however if the observed solution is known to be exactly geostrophic and the Kistler-McPherson correction is used together with a geostrophic adjustment procedure. After geostrophic adjustment has been applied once, the difference flow is and remains geostrophic. The energy convervation argument presented in Section 2 then applies in the particular case of a purely geostrphic flow, and an assimilation will converge under the only condition that the available observations uniquely define the observed geostrophic solution. In the case of a periodic f-plane, it is easy to show that the assimilation will have exactly converged (i.e. the difference flow will have been reduced to exactly 0) as soon as the mass field has been updated at least once at each grid point.

Coming back to the general correction scheme (33), we are now going to study how such a scheme modifies the amplification matrix (10). We will assume that the correction matrix D can vary with the introduction time. After introduction of observations of X, and correction of Y, at time t_i, we are left with differences $\Delta X = 0$ and $\Delta Y(t_i)$. Integrating the model to the next introduction time t_{i+1} will produce the following differences (Eqs. (6)).

$$\Delta_1 X(t_{i+1}) = R^y_x(t_{i+1},t_i) \, \Delta Y(t_i) + o(\Delta Y(t_i))$$

$$\Delta_1 Y(t_{i+1}) = R^y_y(t_{i+1},t_i) \, \Delta Y(t_i) + o(\Delta Y(t_i))$$

Introducing the observations at time t_{i+1}, and correcting Y according to (33) will set ΔX to 0 and transform ΔY into

$$\Delta Y(t_{i+1}) = \Delta_1 Y(t_{i+1}) + D(t_{i+1}) \, \Delta_1 X(t_{i+1}) + o(\Delta_1 Y(t_{i+1}), \Delta_1 X(t_{i+1}))$$

$$= [R^y_y(t_{i+1},t_i) + D(t_{i+1}) \, R^y_x(t_{i+1},t_i)] \, \Delta Y(t_i) + o(\Delta Y(t_i)) \quad (34)$$

The developments of Section 2 therefore remain valid, the matrix $R^y_y(t_{i+1},t_i)$ being replaced at each assimilation step by

$$P^y_y(t_{i+1},t_i) = R^y_y(t_{i+1},t_i) + D(t_{i+1}) \, R^y_x(t_{i+1},t_i)$$

In particular the amplification matrix (10) now becomes

$$A' = P^y_y(\tau_1,\tau_M) \, P^y_y(\tau_M,\tau_{M-1}) \ldots\ldots P^y_y(\tau_2,\tau_1) \quad (35)$$

A correction of type (33) will be useful if the resulting spectral radius $\rho(A')$ is smaller than $\rho(A)$. A particularly interesting case is when the matrix A' can be made equal to 0 by an appropriate choice of the correction matrices $D(t_i)$. If this can be achieved, the decrease of the difference ΔY with the number of assimilation cycles will be faster than exponential. The following theorem is proved in T77.

Theorem (T) : The matrix A' can be made equal to 0 by an appropriate choice of the correction matrices $D(t_i)$ if, and only if, the following condition (C) is satisfied

(C): the only solution of the linearised perturbation Eqns. (3) which satisfies the condition $\delta X(t_i) = 0$ at all observation times t_i (i=1,...,N) is the null solution $\delta x(t) \equiv \delta y(t) \equiv 0$.

Condition (C) essentially means that, in the approximation defined by the linearised Eqns. (3), the available observations $X(t_i)$ uniquely define the observed solution.

The proof of theorem (T) resorts only to basic notions of linear algebra. It turns out that a complete assimilation cycle is not necessary to make the matrix A' equal to 0, but that the product of the matrices P_y^y over either of the two phases (forward or backward) of a cycle can be made null by an appropriate choice of the correction matrices $D(t_i)$. Also, it is not necessary for theorem (T) to hold that the same parameters be observed at the successive observation times, but the nature and even the numbers of the observed parameters can vary with time. Moreover, the set of correction matrices which make A' equal to zero is in general not unique.

Example

Let us consider the linearised shallow-water equations on an f-plane (Eqs. (15)). For successive observations of the mass field ϕ separated by $\Delta\tau$, the matrices R_x^y and R_y^y are obtained from the resolvent matrix (16), viz

$$R_x^y = -\frac{\phi_o}{\alpha} \begin{pmatrix} \sin\beta & \gamma(1-\cos\beta) \end{pmatrix}$$

$$R_y^y = \begin{pmatrix} \cos\beta & \gamma\sin\beta \\ -\gamma\sin\beta & 1-\gamma^2(1-\cos\beta) \end{pmatrix}$$

Given N successive observations of ϕ separated by $\Delta\tau$, we have seen in Section 2 that these observations uniquely define the corresponding solution of (15), i.e. they satisfy the above condition (C) if, and only if, the following conditions are simultaneously verified

$N \geq 3$
$\beta \neq \ell\pi$ ℓ integer
$\gamma \neq 0$
$\gamma \neq 1$

Theorem (T) tells us that, under these same conditions, there exist correction matrices D, with dimensions 2 x 1, which make the matrix A' of (35) equal to 0. Since condition (C) requires only N ≥ 3, two such correction matrices must be sufficient. There must therefore exist two 2 x 1 matrices D_1 and D_2 such that

$$(R_y^y + D_2 R_x^y)(R_y^y + D_1 R_x^y) = 0$$

An easy calculation shows that one solution to this equation is

$$D_1 = D_2 = \frac{\alpha}{2\Phi_o} \begin{pmatrix} \frac{1 + 2\cos\beta}{\sin\beta} \\ \\ \frac{1-2\gamma^2(1-\cos\beta)}{\gamma(1-\cos\beta)} \end{pmatrix}$$

This is always defined, except for values of Φ_o, β, γ for which condition (C) is not satisfied anyway.

Theorem (T) provides a theoretical basis for optimising the convergence of an assimilation. However, the explicit computation of a set of "optimal" correction matrices $D(t_i)$ in an operational assimilation raises a number of difficulties, the most basic of which is the following: the optimal matrices depend in the linearised perturbation Eqns. (3). The latter, in the case when the basic equations (1) are non-linear, depend in turn on the observed solution, which is precisely what is being looked for. It is therefore certainly impossible to determine the correction matrices which make the amplification matrix A' exactly equal to 0. But, since an assimilation performed with meteorological equations already converges with no correction of type (33) at all, it is reasonable to assume that its convergence can be accelerated by optimal matrices corresponding, not to the solution which is actually observed, but to some already known solution which can be considered as being some approximation of the observed solution.

Finally, it must be added that theorem (T) *per se* provides a way of using only what can be called the _dynamical_ information, obtained from the fact that the solution to be determined must satisfy the basic equations (1). It is of no direct help for using any kind of _statistical_ information, obtained from known statistical properties of the solution to be reconstituted. Indeed such statistical information is commonly used in present assimilation procedures under the form, for instance, of structure functions. Using statistical information amounts to reducing the number of independent degrees of freedom to be determined. A fully efficient assimilation procedure must use information of both kinds. A possible way for doing so could be to restrict the linearised perturbation Eqs. (3) to those modes which are compatible with the a priori imposed statistical constraints.

7. ACKNOWLEDGEMENTS

The author thanks several of his colleagues at Laboratoire de Météorologie Dynamique, especially Pr. P. Morel and Dr. R. Sadourny, for numerous stimulating discussions and useful suggestions. The numerical computations were performed on the CDC 7600 computer of Centre National d'Etudes Spatiales, Toulouse, whose collaboration is deeply appreciated. The author is particularly grateful to M.G. Rabreau, who took care of all the programming aspects, and extends his thanks to Iris Rhodes who typed the manuscript, and to Ms. R. Shambrook and M.R. Dufour who drew the figures.

REFERENCES

Coddington, E.A. and Levinson, N. 1955 Theory of Ordinary Differential Equations. McGraw-Hill Book Company, London.

Daley, R. and Puri, K. 1980 Four-Dimensional Data Assimilation and the Slow Manifold. Mon.Wea.Rev., 108, 85-99.

Kistler, R. and McPherson, D. 1975 On the use of a local wind correction technique in four-dimensional data assimilation. Mon.Wea.Rev., 103, 445-449.

Lorenc, A., Rutherford, I. and Larsen, G. 1977 The ECMWF Analysis and Data-Assimilation Scheme: Analysis of Mass and Wind Fields, Technical Report No. 6, ECMWF, Reading, U.K.

Sadourny, R. 1972 Approximations en différences finies des équations de Navier-Stokes appliquées à un écoulement géophysique. Part III. Ann.Géophys., 28, 789-802.

Tadjbakhsh, I.G. 1969 Utilization of time-dependent data in running solution of initial-value problems. J.Appl.Meteor., 8, 389-391.

Talagrand, O. 1977 Contribution à l'assimilation quadri-dimensionelle d'observations météorologiques. Doctoral Thesis, Université Pierre-et-Marie Curie, Paris, France. Available from the author.

Talagrand, O. 1981 On the mathematics of data assimilation. To be published in Tellus.

Varga, R.S. 1962 Matrix iterative analysis. Prentice-Hall, Inc. Englewood Cliffs, New Jersey, USA.

Williamson, D. and Dickinson, R. 1972 Periodic updating of meteorological variables. J.Atm.Sci., 29, 190-193.

SOME CLIMATOLOGICAL AND ENERGY BUDGET CALCULATIONS
USING THE FGGE III-b ANALYSES DURING JANUARY 1979

Masao Kanamitsu[*]

ECMWF

ABSTRACT

The analyses (level III-b) of complete data sets (level II-b) of the First GARP Global Experiment during January 1979 at the European Centre for Medium Range Weather Forecasts are examined from climatological and diagnostics points of view. Comparison with long time averaged climatology is performed. Some difference are discussed in relation to the data coverage.

The tropical velocity potential field for the monthly mean wind is presented. Effects of the initial guess and the initialization are discussed

The kinetic energy budget in zonal wavenumber domain is computed, and results are compared with other studies. Interesting differences in energetics between the hemispheres are found. It is shown that the energetics of the ultralong waves have large latitudinal and temporation variations in the northern hemisphere.

1. INTRODUCTION

The major objective of the meteorological data analysis is to obtain the values of the meteorological variables at regularly spaced grid points from randomly spaced observations with different accuracies and correlations under various mathematical and physical constraints. This objective can be accomplished without serious difficulties by the use of least square interpolation methods under statistical and physical constraints only if the observations are reasonably evenly distributed and the physical constraints are of a simple form. In order to relax those conditions which cannot be met in the case of meteorological analysis, the use of an 'initial guess field' has been introduced - the idea being that, by modifying the interpolation of observations themselves to the interpolation of the deviations from the first guess field, we can utilise the guess field where observations are very sparse or are not very reliable. In addition, if we can introduce an initial guess field which satisfies complex physical constraints as in the atmosphere, it is much easier to design the interpolation of increments scheme which can also satisfy them. It is easily recognised from this point of view that the use of physically consistent and reliable initial guess fields becomes one of the most essential parts of the analysis scheme.

---

[*] On leave from Electronic Computation Centre, Japan Meteorological Agency

It should be emphasised here that, when initial guess fields are introduced, the final analysis field is composed of a mixture of information from observations (with some statistical and physical constraints) and from initial guess fields. The degree of dependency of the analysis to the initial guess field is a complex function of the distribution of observations, the physical constraints applied and the error (observation and prediction) statistics used in the interpolation scheme.

The dependency of the analyses to the initial guess field becomes more important when the analysed fields are used as initial states for numerical forecasts. In this case, it is desirable that the initial guess fields are dynamically consistent with the particular prediction model to be used. This is particularly important for minimising the changes resulting from initialization. For this reason, prediction by the forecast model becomes the first choice as the guess field.

The use of the predicted fields as initial guesses is, thus, very logical and reasonable but one must realise that the predicted fields do have errors inherent in the prediction model itself. It is possible that the errors in the initial guess fields (produced by the model) may accumulate as the analysis-forecast cycle continues, creating undesirable errors and biases in the analyses.

It is the major objective of this study to investigate such errors and to examine the degree of dependency of the analyses to the model. To do this, we first compare the monthly means of analyses with normal climatologies to find any systematic errors. Then, we will perform several diagnostic calculations to examine physical and dynamical consistencies. We feel this is particularly important for the FGGE analyses (III-b data) which are to be used for various climatological and diagnostic studies.

Another important objective of this paper is to examine the impact of the FGGE observation systems in the study of climatology and the general circulations. Although examinations of the analysis system and application of its products may not be discussed at the same time, the large amount of FGGE observations allows us to do so because the effect of the observations is considered to override minor deficiencies of the analysis scheme.

2. ANALYSIS OF THE FGGE DATA AT ECMWF

The analysis scheme used for the production of III-b data is essentially the same as that of the operational system, Lorenc (1980). Major differences are the introduction of observation errors for various FGGE observing systems and some modification of the vertical correlation functions which have smoother variations with height. Some of the detail of the assigned observational errors are given in the article by Bengtsson and Källberg (1980).

The analyses are performed at the standard pressure levels (10, 20, 30, 50, 70, 100, 150, 200, 250, 300, 400, 500, 700, 850, 1000mb) with a horizontal resolution of $1.875°$ on a regular latitude/longitude grid. There are two constraints applied in relation to the analysis increments, i.e. (1) geostrophic balance between height and wind (not an exact balance and relaxed towards equator) and (2) local non-divergence of the wind on scales less than about 660km. Since these constraints are somewhat crude, interpolated increments cannot maintain the dynamical balance established in the initial guess field by the prediction model. To re-establish this balance, non linear normal mode initialization (Temperton and Williamson, 1979) is applied before the forecast is made.

During the analysis - initialization - prediction cycle, several different types of fields become available, i.e. the raw analysis field, the initialised field, and the initial guess field. In addition, the fields are available on pressure levels and on σ-levels due to the use of the σ-coordinate system in the forecast (and, naturally, in the initialization). It is noted that the conversions of fields from pressure to sigma and sigma to pressure surfaces involve vertical interpolation/extrapolation procedures which can introduce inaccuracies. The various fields produced and the variables available are summarised in Table 1.

Table 1   Fields produced during analysis cycle

type of field	vertical coordinate	vertical interpolation	variables
(1) analysis	p	no	u,v,z,q
(2) analysis	σ	yes	"
(3) initialized	σ	no	u,v,z,T,ω,q
(3) initialized	p	yes	"
(4) initial guess	σ	no	"
(5) initial guess	p	yes	"

The steps of an analysis cycle and available variables at each stage may be summarised as follows:

1. Statistical interpolation of increments.

In this step, it is noted that the variables analysed in the Centre's multi-variate statistical interpolation method are the wind components (u and v) and height (z).

Temperature observations are converted to thickness when analysing the height field; no temperature analyses are made. Moisture is separately analysed in the form of layer mean precipitable water content by the use of the one scan correction method.

2. Interpolation from pressure to sigma levels.

The cubic spline interpolation method is used for this interpolation. At this step, the topography used by the prediction model is introduced to help determine sigma levels. All the necessary derivable quantities, e.g. temperatures and divergences are evaluated using vertical and horizontal finite differencing schemes compatible with the prediction model.

3. Nonlinear normal mode initialization.

This procedure modifies the variables u, v, z, p_s to establish proper balance between the motion and the mass fields. It should be noted that, at this point, all the variables u, v, z, p_s and the derived variables T and ω are physically consistent and dynamically balanced (in terms of the initialization procedure applied).

4. Six hour forecast.

The forecast is made with the model with full physics.

5. Interpolation (and extrapolation) from sigma to pressure levels.

This provides initial guess fields for the subsequent analysis.

6. Back to step 1.

There are extra steps associated with step 3, that is the interpolation from sigma to pressure levels. The purpose of this step is to check the effect of initialization on the initial analyses field. It does not affect the analysis cycle at all since this is an independent step. It is also noted that the interpolation in the vertical necessary in this step does affect the physical consistencies and the dynamical balance established by the initialization in the σ-coordinates.

For the general use of the analyses, it is desirable that the analyses are free from any influence of a particular model, e.g. vertical and horizontal grid structures, finite differencings, smoothed model topographics and especially model physics. For this reason, the FGGE III-b data produced at the Centre includes analysed (raw, uninitialized) u, v, z, p_s which are free from model influence (except by the initial guess field). They also include initialized T and ω at the pressure levels for numerical forecast uses, which are strongly dependent on the Centre's model. It is to be noted that the u, v, z, p_s and T, ω are not physically consistent, i.e. thickness of analysed heights does not correspond to T, and ω does not correspond to the vertically integrated divergence of the analysed winds.

The difference between the analysed divergence and the initialized divergence will be discussed in detail later. However, it must be emphasised that the latter is strongly dependent on the Centre's model structures and the method of initialization and, accordingly, may not represent the divergence in the real atmosphere. The user of the III-b analysis is advised to be aware of this fact.

In this paper, since we are mostly interested in the climatological and diagnostic aspect of the analysis, only raw analysis fields (u, v and z) are considered. The derivable quantities (e.g. T, etc.) are computed directly from the raw analysis variables (without going through p to σ transformation) so that no initialized fields are used. Furthermore, in order to minimise the effect of the initial guess fields, the analyses at 00 and 12 GMT are used to maximise the data coverage. The monthly averaged field shown in this paper is, thus, computed from twice daily analyses (62 analyses in January).

3. EXAMINATION OF MONTHLY MEAN FIELDS

The observational network during FGGE improved considerably the data coverage in the tropics and southern hemisphere. The enhanced cloud winds from five satellites covered nearly the entire globe from $50^{o}S$ to $50^{o}N$ providing wind information in both the lower and upper troposphere. The drifting buoys in the southern hemisphere oceans provided high quality pressure information and aircraft, constant level balloons, special observing ships, etc. supplied essential information in the tropics. An example of the distribution of FGGE observations can be found in the preface to this book.

Due to the improvement of observing systems, the monthly climatology may be considerably different from 'old climatology' based on conventional observational coverage, especially in the tropics and the southern hemisphere. The comparison of the monthly mean with 'old climatologies' cannot provide accurate information on the systematic errors and biases of the analysis scheme in these areas; rather, it provides a measure of the impact of the FGGE observations on the deduced monthly climatology. Even in the data dense northern hemisphere, annual variation which is known to be very large, can hinder identification of the deficiencies of the analyses. In this sense, we must recognise that only major deficiencies can be detected by this comparison.

Two sources of 'old climatology' are selected in this study. For the high latitudes, a long time monthly average used at ECMWF (originated from the NCAR data archive) is used. For the tropics and for the global wind statistics, seasonal climatology by Newell et al (1972) is chosen.

This climatology is different from the conventional ones in the
sense that it is based on the time averaged observations and not on
the time averaged analyses, so that no influence of analysis is
involved except at the final analysis stage. In the seasonal
climatology maps, the values at the stations are all plotted and the
effect of the final analysis stage can be recognised very easily.

In this paper, we discuss the monthly average of January only, due
mainly to the availability of the analyses at the time of performing
the actual calculations. Further evaluations of other months (up
to July) have been performed since then and will be published by
ECMWF soon.

In the following, we discuss synoptic and some diagnostic
comparisons of the FGGE and the 'old climatologies'.

3.1   Northern hemisphere

3.1.1 Surface pressure (Fig. 1)

Several large deviations from normal are noted as follows:

(i)   Very weak Icelandic low
(ii) Intense Aleutian low
(iii) Intense Siberian high
(iv) Weak subtropical highs

These anomalous states over the high latitudes are the results of
intense blocking which persisted over Alaska (first half of the
month) and over Greenland (last half of the month).

3.1.2 500mb height (Fig. 2)

As a result of the blockings, the ridges over Alaska and northern
Atlantic are very intense.

These anomalous states during January 1979 make it somewhat difficult
to detect systematic errors in the analysis scheme.

3.2   Southern hemisphere

We note here again that, in the southern hemisphere and in the
tropics, the deviation from old climatology is due mainly to the
increased observations and not to the analysis scheme.

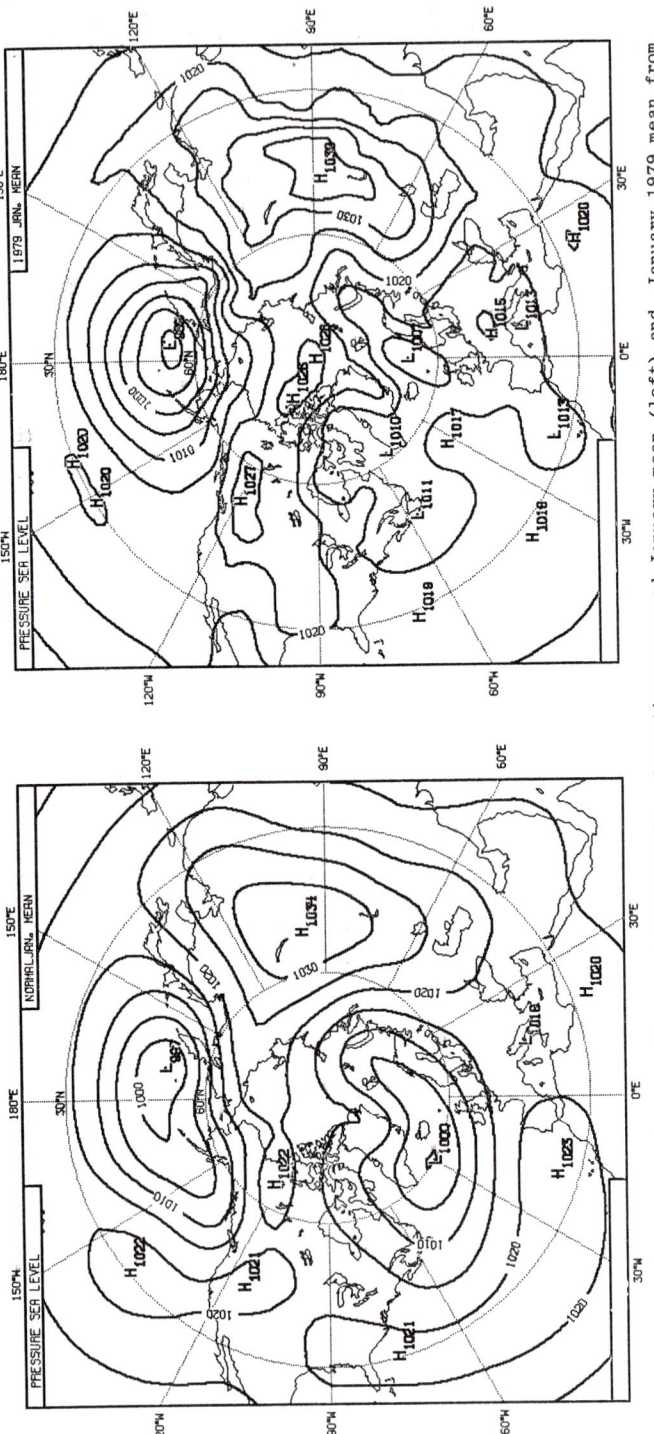

Fig. 1 Northern hemisphere surface pressure pattern. Long time averaged January mean (left) and January 1979 mean from ECMWF analyses (right). Units in mb. Contour interval is 5 mb.

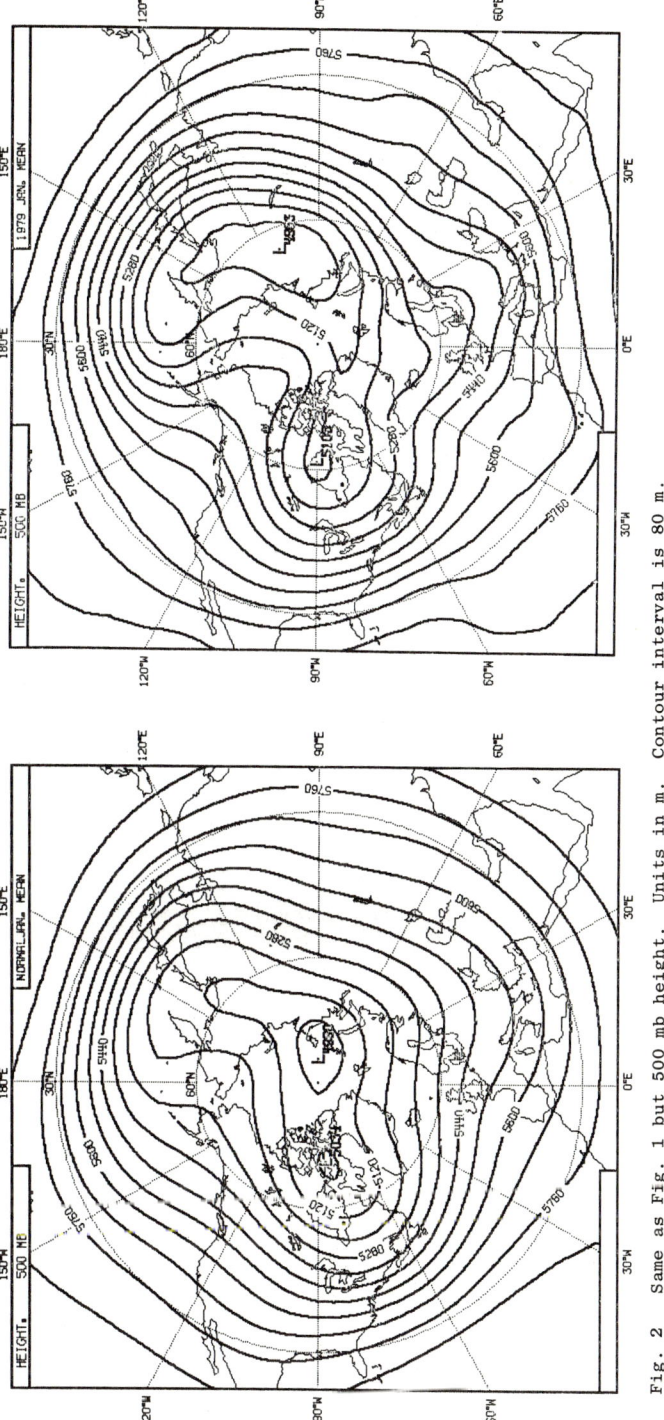

Fig. 2 Same as Fig. 1 but 500 mb height. Units in m. Contour interval is 80 m.

3.2.1 Surface pressure (Fig. 3)

The lows located around $60°S$ are 3-4mb lower than the old climatology. The subtropical highs are higher by 2-4mb and, as a result, the pressure gradient in the middle latitude zone is larger. The high located off the east coast of South America ($40°S$, $45°W$) has not been depicted in the old climatology.

3.2.2 500 mb geopotential height (Fig. 4)

Zonal asymmetry is more pronounced than in the (old) climatology; specifically, the troughs around $40°W$ and $90°W$ are very distinct. Dominance of zonal wavenumber 5 is noted, which is also clear in the surface pressure field (Fig. 3).

3.3 Tropics

3.3.1 850 mb wind field

In general, the major large scale features are similar to climatology (Fig. 5). The following differences are noted.

(i) Larger equatorial westerly zone (extending from $60°E$ to $160°W$).

(ii) Distinct trough located off the east coast of South America, separating the south Atlantic subtropical high into two.

(iii) Weaker subtropical highs in the northern hemisphere and the trough over the west coast of Africa.

Fig. 6 shows that zonal winds are much larger than in Newell's mean. The differences are largest over the equatorial eastern Pacific and the equatorial Atlantic, where conventional data coverage is very poor.

A strong equatorial westerly jet (~15 m/sec) exists in the area between Australia and New Guinea. The jet is strongest at 850 mb and becomes weaker above and below. This jet is observed only in January and is much weaker in February.

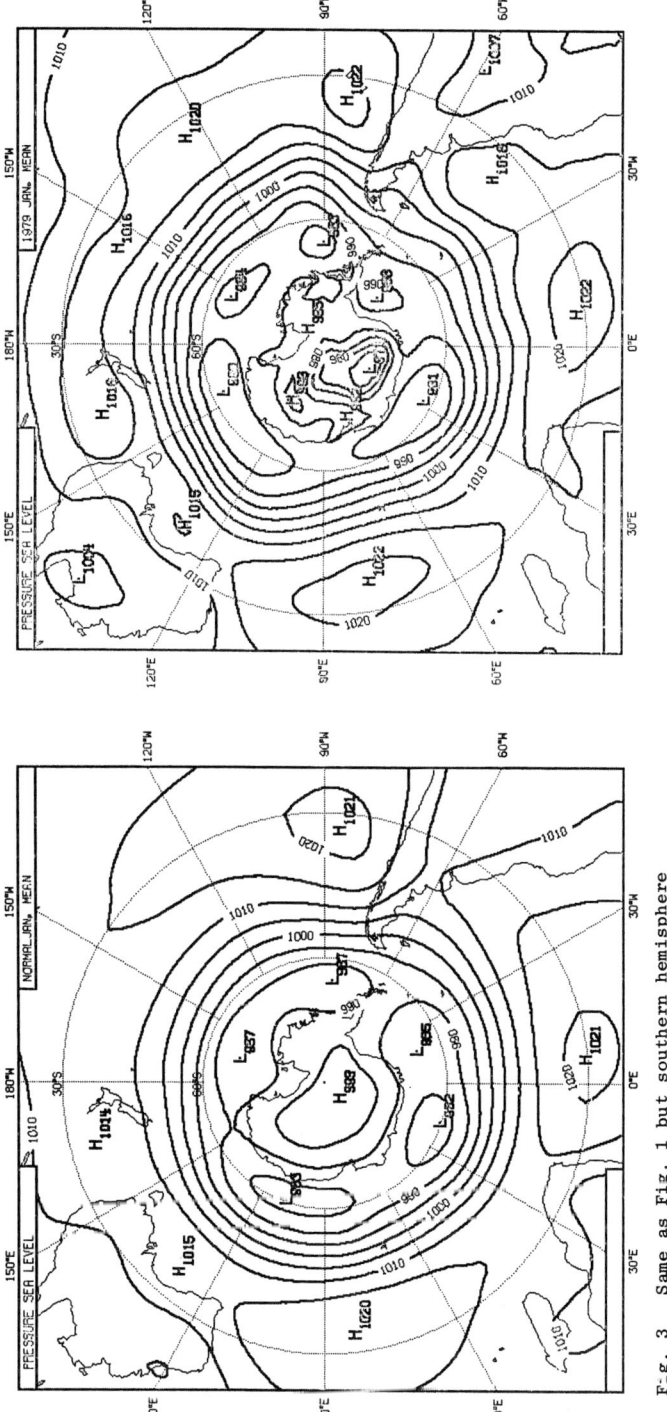

Fig. 3 Same as Fig. 1 but southern hemisphere

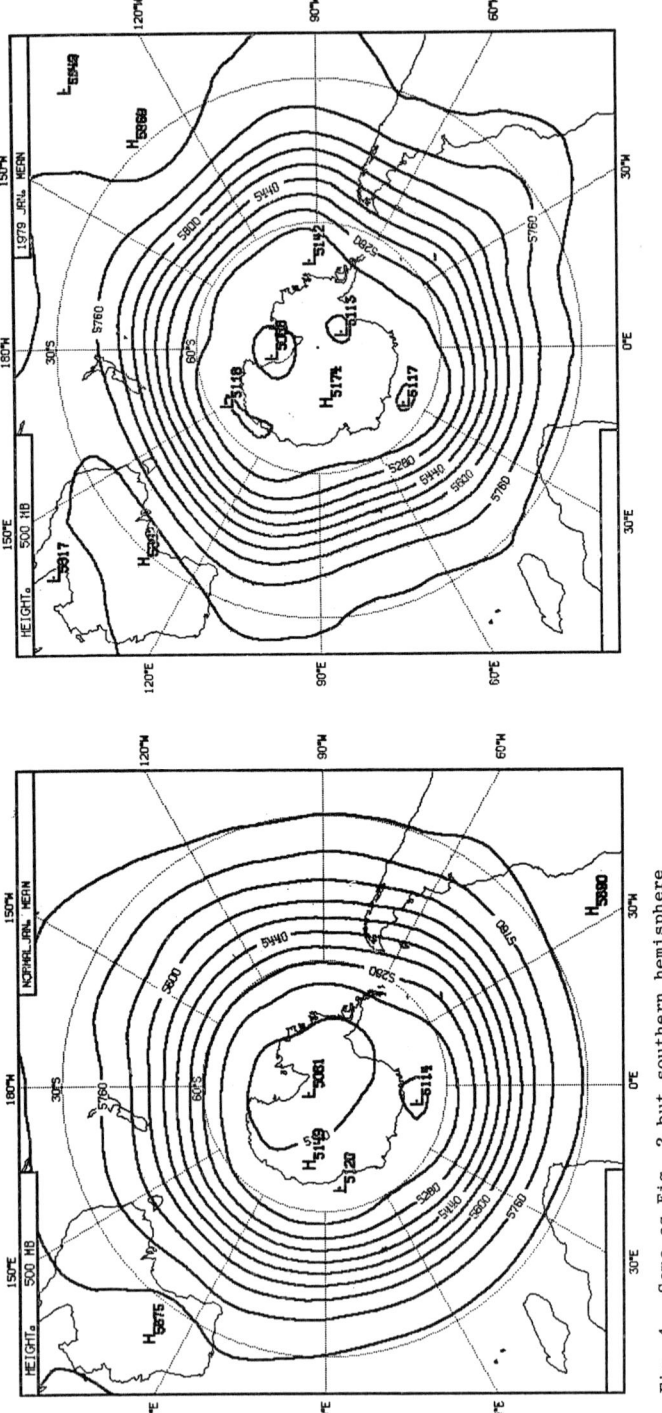

Fig. 4 Same as Fig. 2 but southern hemisphere

Fig. 5 850 mb streamlines. Winter average in the tropics by Newell et al, 1972 (top). ECMWF January mean (bottom) over the globe.

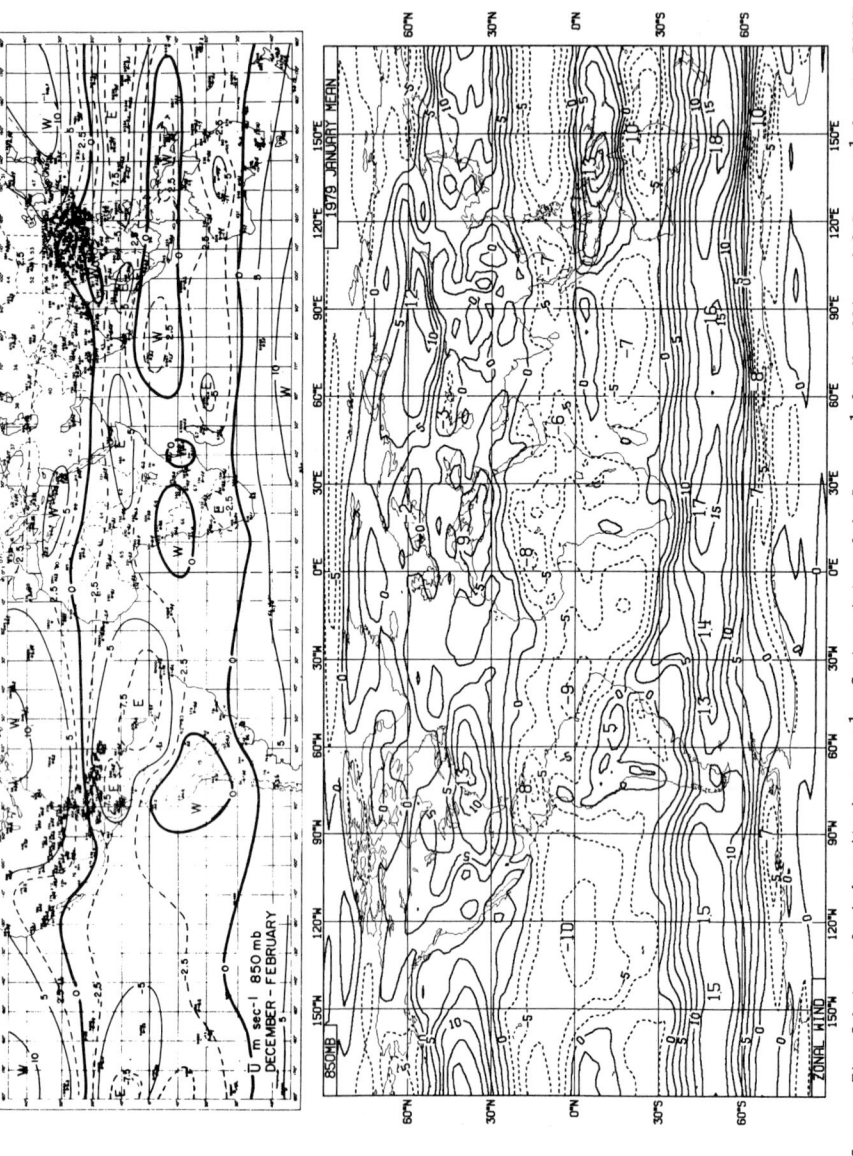

Fig. 6 Same as Fig. 5 but zonal wind, units in m sec⁻¹. Contour interval is 5 m sec⁻¹ for Newell's and 2.5 m sec⁻¹ for the ECMWF analysis. In the bottom figure, isolines for the negative values are indicated by dashed lines. In the top figure dashed lines indicate intermediate contour levels (2.5 m sec⁻¹).

Magnitudes of the meridional wind of the ECMWF analysis (Fig. 7) are also larger than the climatology. The largest differences are observed over ocean areas, particularly over the southern hemisphere and Indian ocean.

Three areas of strong cross equatorial northerlies are noted, over the Amazon basin, the east coast of Africa (in a very narrow zone, somewhat resembling the Somali jet in the summer monsoon season but reversed in direction) and Indonesia. The westerly jet over southern New Guinea corresponds to the zone of strong convergence in the meridional wind field.

3.3.2 200 mb wind field

Major flow patterns at this level (Fig. 8) are also very similar to Newell's. Some differences are again observed over the ocean areas.

(i) The trough over Mexico extends to the southern hemisphere ($100°W$).

(ii) A very sharp trough with a strong tilt from north west to south east is located at $140° - 120°W$, $40°S$ - equator.

For the zonal winds (Fig. 9), the area of easterlies in the tropics is somewhat smaller in the ECMWF analysis and mostly located over the continents (corresponding to the northern part of anticyclonic vortices located over the continents). The magnitudes are smaller than in Newell's analysis. The pattern over the south-eastern Pacific shows interesting detail with weak easterlies at $30°S$, a westerly jet at $15°S$ (corresponds to the trough extending from Mexico), and a weak westerly zone at $10°N$ separating the jet from the subtropical jet at $30°N$.

Meridional winds (Fig. 10) at 200 mb are 2-3 times larger than in Newell's analysis, and differences are again large over the southern hemisphere ocean areas. Two zones (tilted from southwest to north east) of southerly cross-equatorial flow exist off the west and east coast of South America. Very strong cross equatorial southerlies are shown in a film of infrared cloud imagery during the special observation period over a similar area (Fleming and Bohan, 1980, presented at the International Conference on Preliminary FGGE Data Analysis and Results, Bergen, Norway).

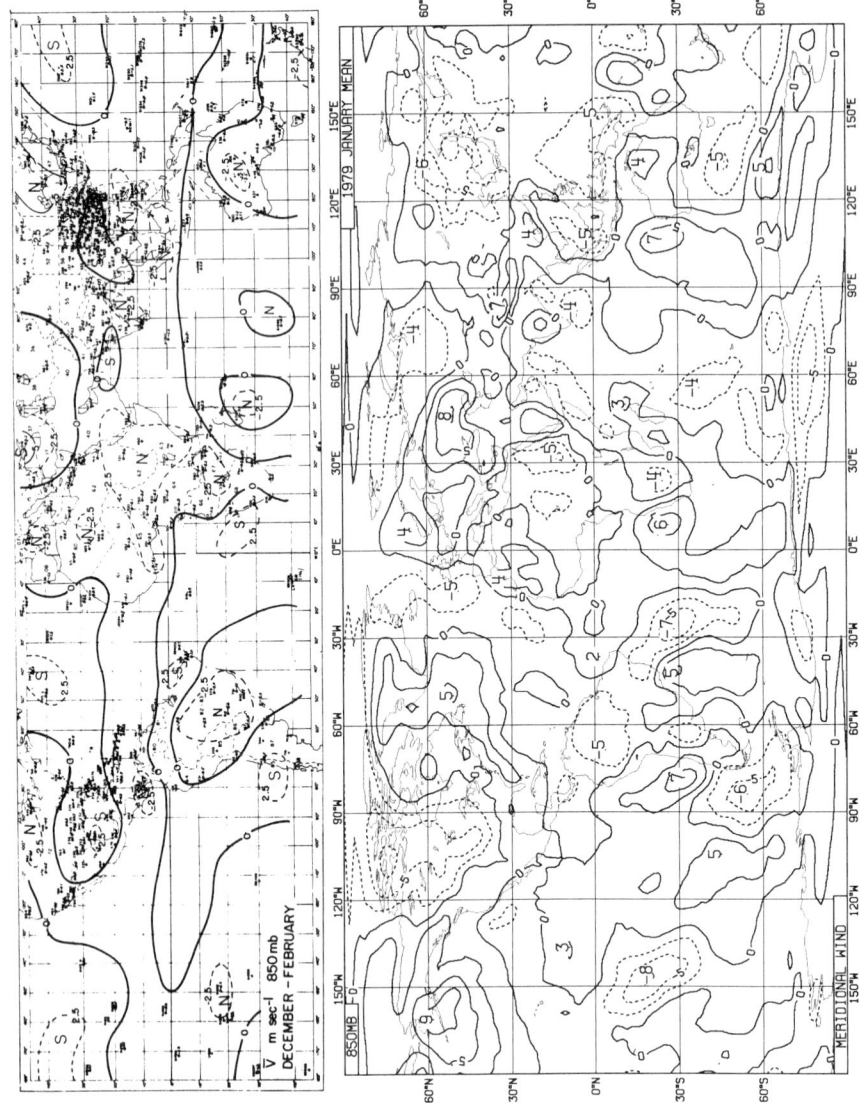

Fig. 7 Same as Fig. 6 but meridional wind at 850 mb

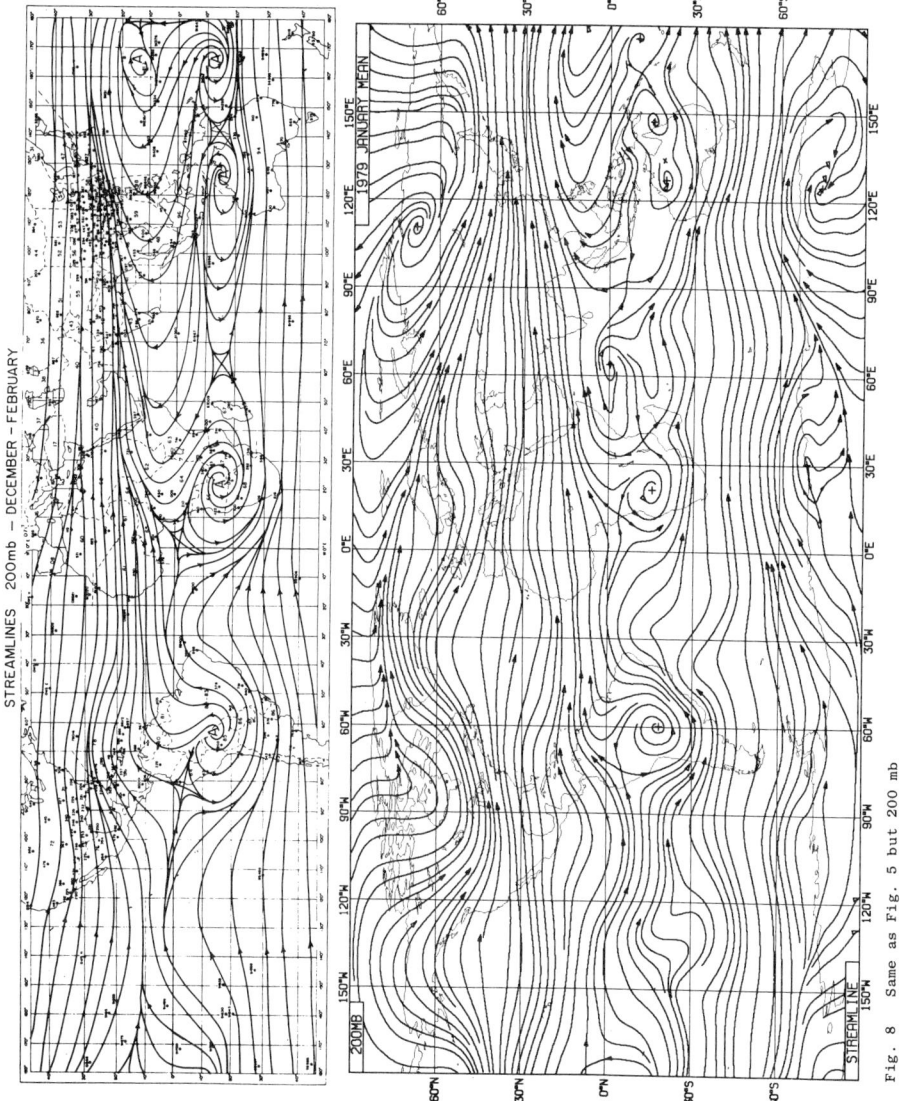

Fig. 8 Same as Fig. 5 but 200 mb

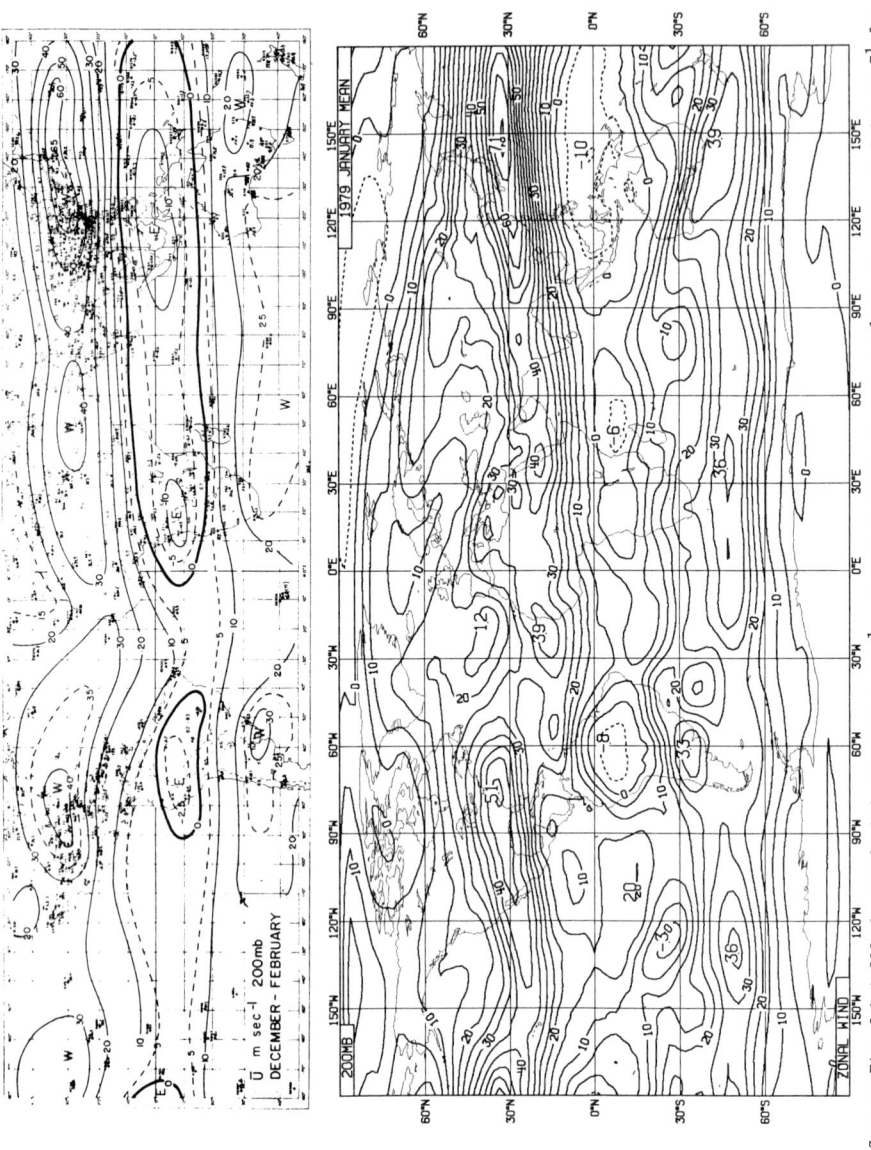

Fig. 9 Same as Fig. 6 but 200 mb, zonal wind, units in m sec^{-1}. Contour interval is 10 m sec^{-1} for Newell's and 5.0 m sec^{-1} for the ECMWF analysis.

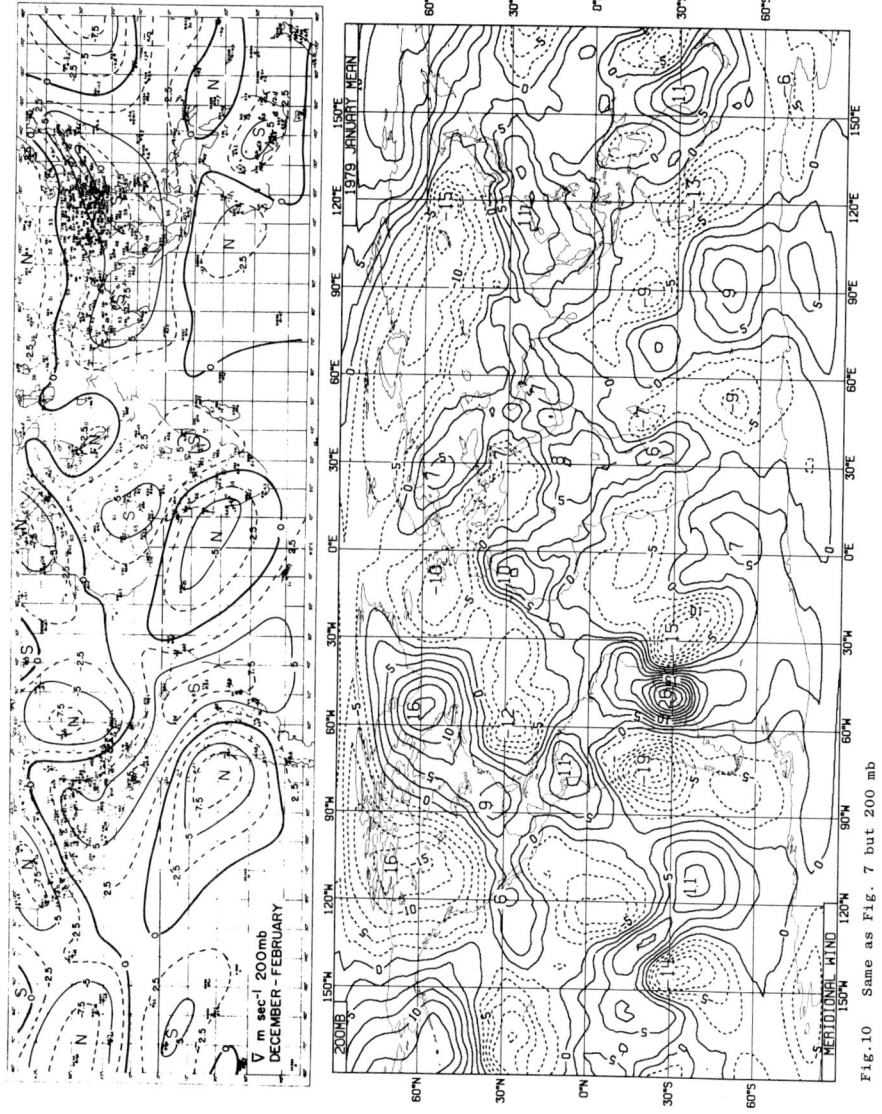

Fig.10 Same as Fig. 7 but 200 mb

3.4 Zonally averaged zonal motion field

Fig. 11 shows the latitude-height cross section of zonally averaged zonal wind, both for the ECMWF analysis and for Newell's.
In the northern hemisphere, differences between the two are very small. In contrast, the westerly jet in the southern hemisphere is about 8 m/s stronger in the ECMWF analysis and the westerly area extends much higher into the stratosphere.

The tropical easterly belt in the ECMWF analysis is observed only below 600 mb and above 100 mb. The mean zonal wind in the middle to upper troposphere is a weak westerly.

3.5 Time variability of u and v

To examine time variability, standard deviations of wind components from monthly averages are computed. Only the zonal mean is shown here. The standard deviations of u and v (Fig. 12) in the northern hemisphere for the ECMWF and Newell's analyses are again not too different. Large differences are observed in the southern hemisphere where the standard deviation of v in the ECMWF analysis is much larger, indicating more amplitude in transient disturbances. On the other hand, time variability of the winds in the tropical upper troposphere is smaller (especially in u-field) in the ECMWF analysis. The cause of this smaller variability is not clear at this stage but it can be related to the difference in the types of observations.

3.6 Momentum transport

Following Newell's analysis, zonal mean momentum fluxes are computed for the ECMWF analysis. The transport is separated into three components, i.e.

$$[\overline{uv}] = [\overline{u}][\overline{v}] + [\overline{u^* v^*}] + [\overline{u'v'}] \qquad (1)$$

The square bracket and over bar stand for the zonal and time average respectively, and star and prime are the deviations from the corresponding means. The three components are named as the fluxes by the mean meridional motion (the first term on the right of Eq. 1), by the standing eddies (2nd term) and by the transient eddies (3rd term).

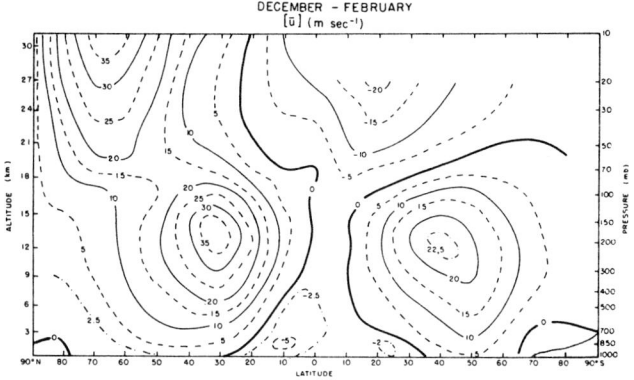

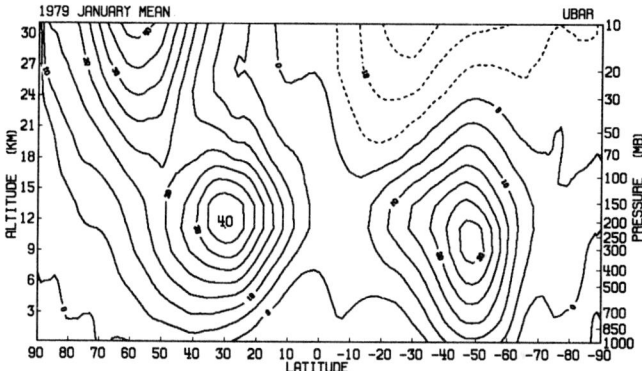

Fig.11 Meridional cross section of the zonally averaged zonal wind. Winter average by Newell et al (top) and January mean from ECMWF analyses (bottom). Units in m sec^{-1}. Contour interval 5 m sec^{-1}.

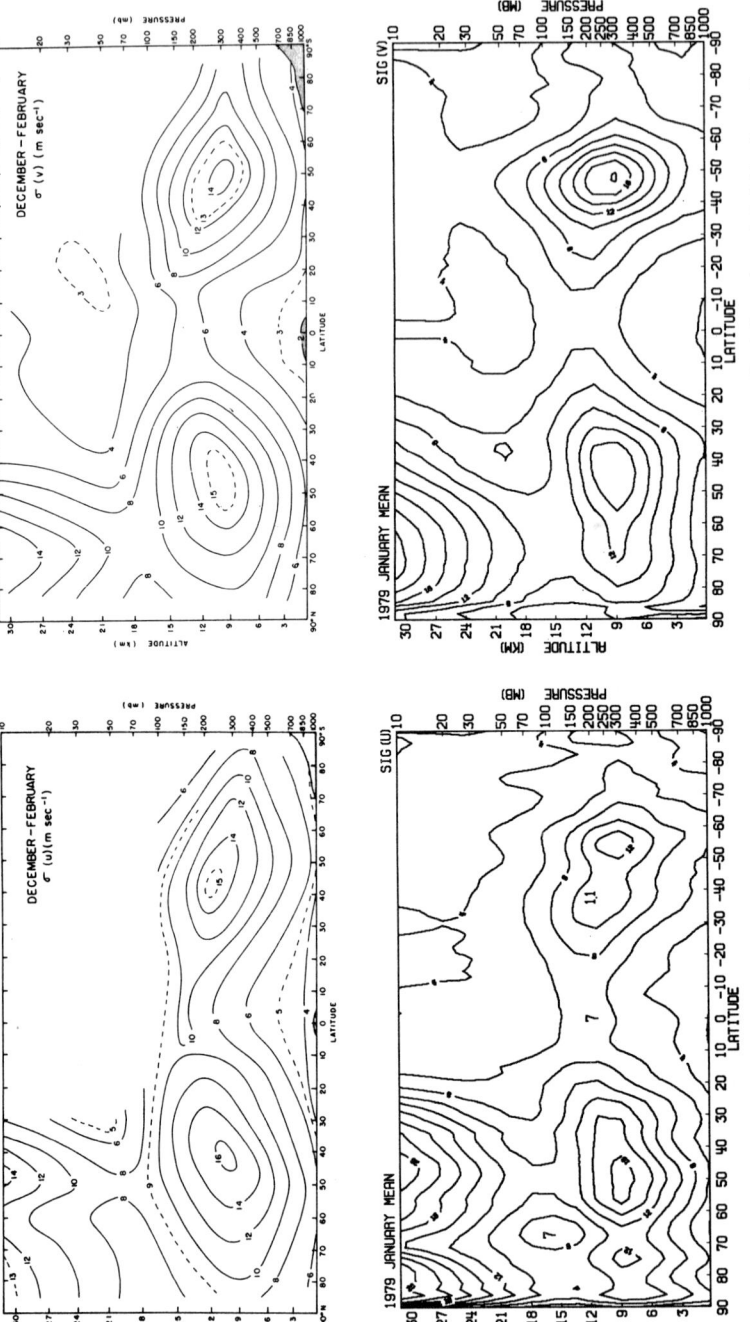

Fig. 12 Same as Fig. 11 but averaged standard deviation of the wind components, u (left) and v (right) from January mean, units in m sec^{-1}. Contour interval is 2 m sec^{-1}.

The momentum fluxes by the mean meridional circulation (Fig. 13 left) show similar patterns for the two analyses. The differences are the stronger fluxes in the southern hemisphere for the ECMWF analysis, particularly between 40°S and 60°S and in the northern hemisphere stratosphere. In the ECMWF analyses, combinations of persistence and climatology are used as the initial guess for the analyses above 50 mb. Since no predicted fields are used, dynamic control of the initial guess field is considered to be somewhat limited. In order to make definite comment on this difference, further investigation of the use of persistence is required.

The significant difference of the momentum flux by the standing eddies (Fig. 13, right) in the northern hemisphere at 60°N may be explained by the blocking. It can also be related to the pronounced stratospheric sudden warming events in the second half of January. The peaks of the northward fluxes at 30°N are similar. Winter to summer hemisphere transport of momentum at the equatorial upper troposphere exists in both analyses but is somewhat weaker in the ECMWF analysis. In the southern hemisphere, southward flux at 30°S stands out in the ECMWF analysis but the flux is smaller in other areas, indicating the quite different character of standing waves between the two analyses.

The momentum flux by the transient eddies (Fig. 14) shows interesting differences at the equatorial upper troposphere, where the local maximum of southward transport of momentum in Newell's analysis is missing in the ECMWF analysis. To examine the difference of the momentum flux in the tropics, the distributions of $\overline{u'v'}$ at 200 mb are compared (Fig. 15). Fair agreement of the signs and magnitudes of the fluxes between the two analyses can be seen in the area where the observations exist in Newell's analysis. Major discrepancies are located over the complete data void area in Newell's analysis (equatorial Atlantic and Pacific) where the fluxes are generally directed towards north. In other areas, the momentum fluxes by the transient disturbances are alike in the two analyses.

3.7 Divergent part of the wind

In order to examine forcing in the tropics, the velocity potential field has been shown to be very useful (Krishnamurti, 1971). The velocity potential (χ) is defined as,

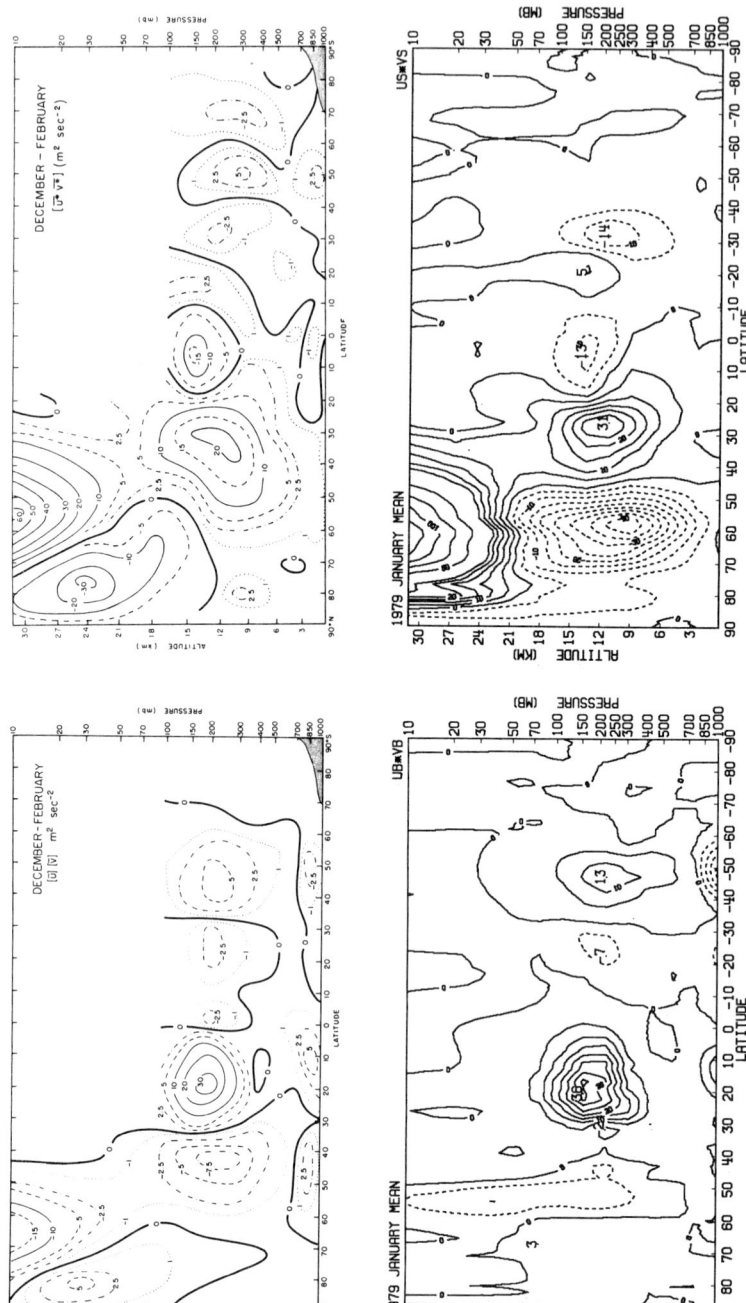

Fig. 13 Same as Fig. 11 but zonally averaged momentum flux by the mean meridional circulation (left) and by the standing eddies (right). Units in m² sec⁻². Contour interval is 5 m² sec⁻².

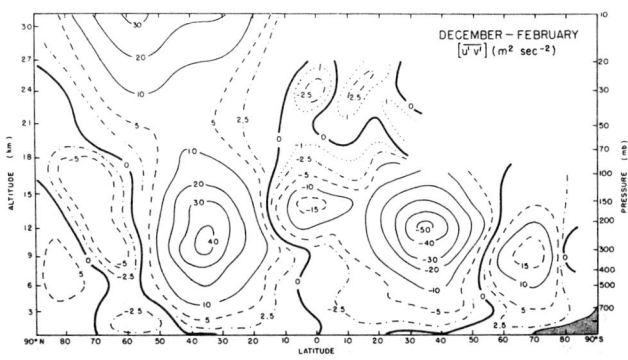

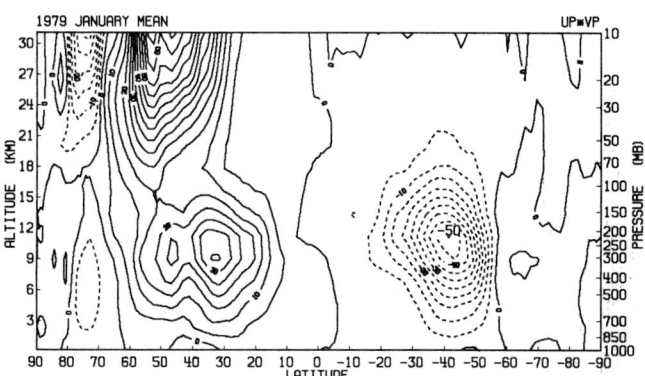

Fig.14 Same as Fig.13 but zonally averaged momentum flux by the transient eddies.

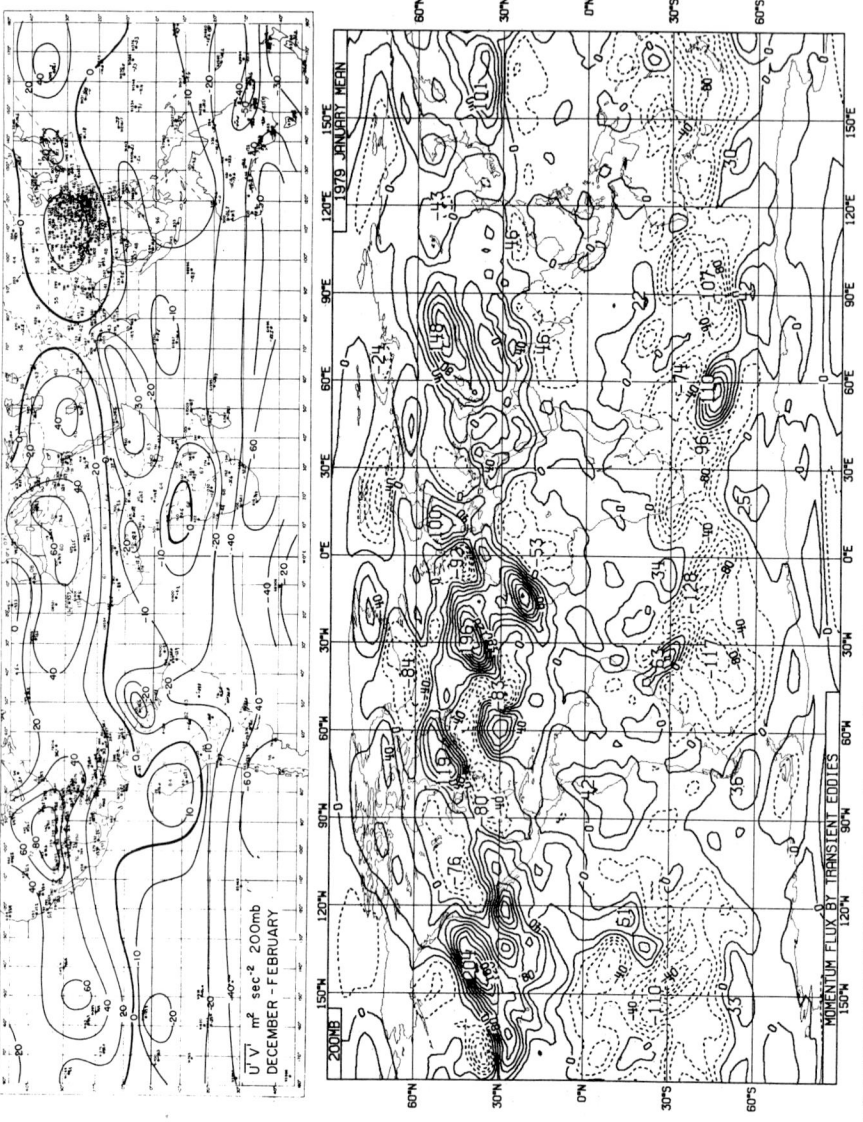

Fig. 15 Horizontal distribution of momentum flux by the transient eddies at 200 mb. Winter mean tropics by Newell et al (top) and January 1979 mean global from ECMWF analyses. Units in m² sec⁻², contour interval 10 m² sec⁻².

$$\nabla^2 \chi = \frac{1}{a \cos\phi} \left(\frac{\partial u}{\partial \lambda} + \frac{\partial v \cos\phi}{\partial \phi} \right) \qquad (2)$$

Here, a is the radius of the earth, ϕ is latitude and λ is longitude. ∇^2 is the Laplacian operator;

$$\nabla^2 = \frac{1}{a^2 \cos\phi} \left\{ \frac{\partial}{\partial \lambda} \left(\frac{1}{\cos\phi} \frac{\partial}{\partial \lambda} \right) + \frac{\partial}{\partial \phi} \left(\cos\phi \frac{\partial}{\partial \phi} \right) \right\}$$

3.7.1 Analysis of divergence and the effect of initialization

As has been discussed in section 2, ECMWF's analysis scheme performs space interpolation of increments (difference between the initial guess field and the observation) from observations to grid points under some constraints. The final raw analysis is obtained by adding the interpolated increments to the initial guess field at the grid points. Therefore, any quantities linearly related to the analysis variables (u, v, z, p_s) are expressed as the sum of the quantities of the increments and of the initial guess field. Thus, the divergence computed from raw analysis of u and v is equivalent to the sum of the divergence of analysis increment and the divergence of the initial guess field. Thus, the divergence computed from raw analysis of u and v is equivalent to the sum of the divergence of analysis increment and the divergence of the initial guess field.

Since the analysis scheme has a constraint that the increments are locally non divergent (i.e. on a scale of less than 660 km), the final divergence field is consistent with that of the forecast model generated divergence (also affected by the initialization, 6 hours ago) and the divergence associated with observations but with a scale greater than 660 km.

For the purposes of evaluating the ability of the analysis scheme to analyse divergences from observations which are normally considered to have large errors, it is essential to examine the contributions of these two sources of divergence. Fig. 16 shows an example of the velocity potential field (χ) at 200 mb (17 January 12 GMT 1979, this date being arbitrarily chosen). The velocity potential of the analysis increment (Fig. 17, upper) computed as a difference of the χ field between the analysis and the initial guess field shows that the observational increments make a

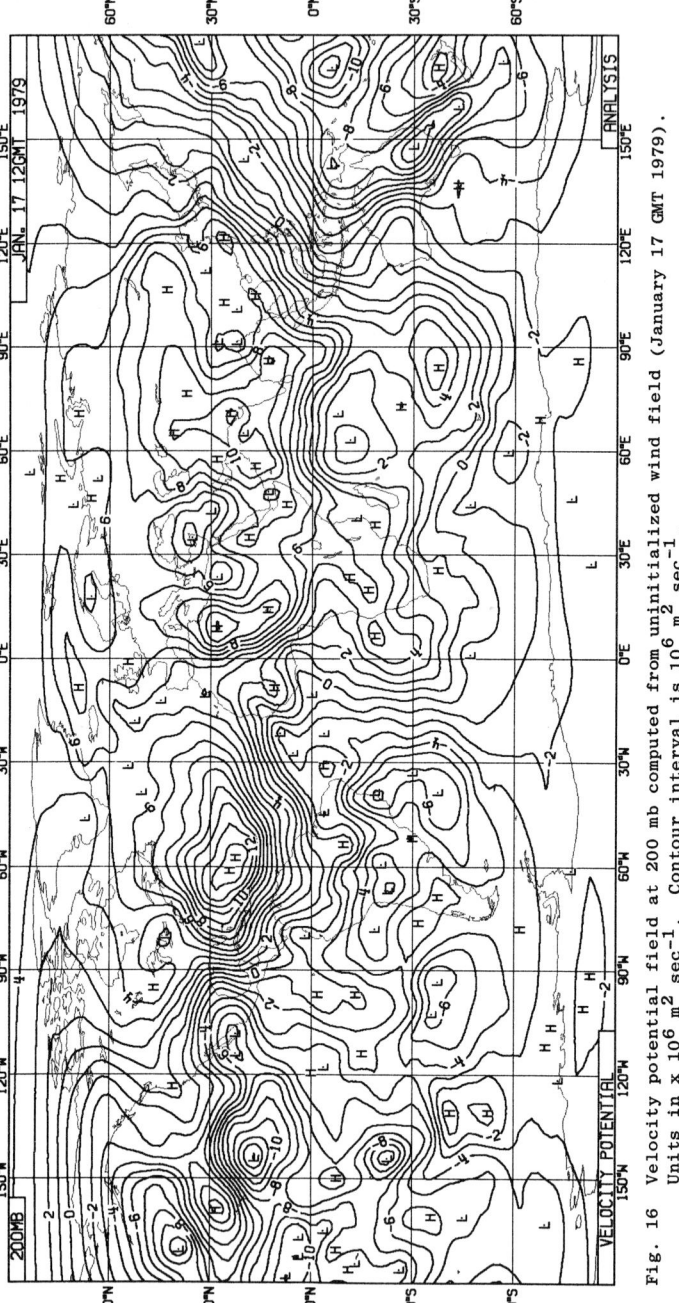

Fig. 16 Velocity potential field at 200 mb computed from uninitialized wind field (January 17 GMT 1979). Units in x 10^6 m² sec⁻¹. Contour interval is 10^6 m² sec⁻¹.

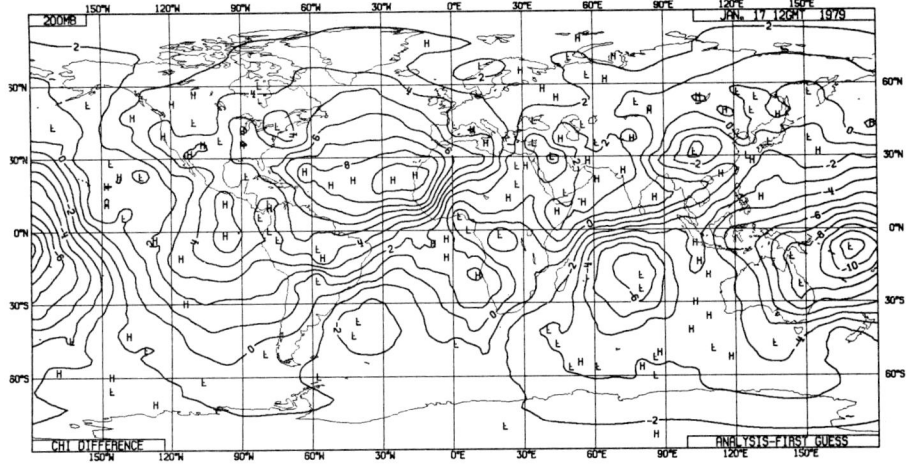

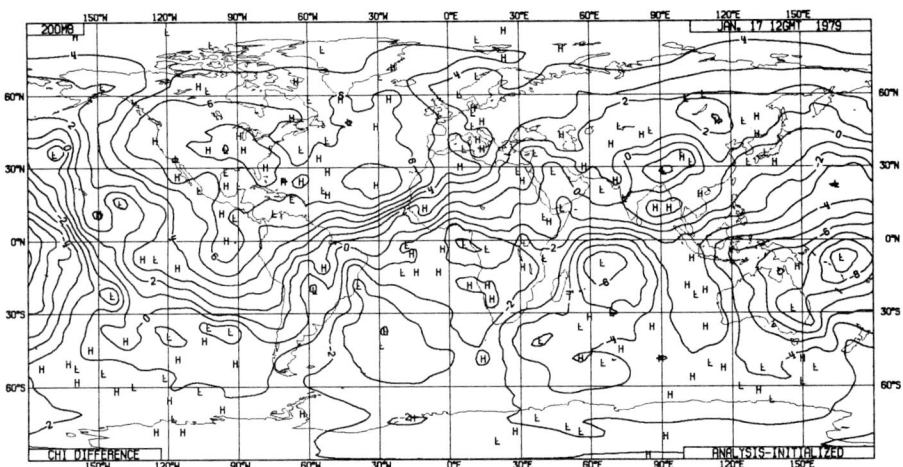

Fig. 17 Difference of velocity potential between the analysis and initial guess (top) and between the analysis and the initialized field (bottom), units in x 10^6 m^2 sec^{-1}. Contour interval is 10^6 m^2 sec^{-1}.

considerable contribution to the final analysed χ field, particularly over the tropics. The similarity of the pattern between the two figures (Fig. 16 and Fig. 17, upper) indicates that the initial guess field has significantly less divergence in the tropics.

This agrees with the findings of the Centre's diagnostic group that the Hadley circulation (zonally averaged divergent part of the wind) is too weak in the forecast, particularly at the initial time. To confirm this, the difference of the field between the analysis and the initialized field is computed and is shown in the lower panel of Fig. 17. It is evident that the analysed divergence is very much reduced by the initialization. The cause of this reduction is considered to be due to the lack of diabatic heating in the initialization but the real reason is unclear. The lack of divergence in the initial guess field results in the slow build up of the diabatic heating in the prediction model for the first 12 to 24 hours and, hence, the model fails to recover the proper magnitude of divergence for the initial guess field.

It should be emphasised here that the initialized divergence is very strongly influenced by the method of initialization and does not necessarily provide true information. This may be particularly so in low latitudes where the divergence is predominantly determined by the diabatic heating.

Based on these discussions, we have utilised the raw analysis values of u, v, z and p_s throughout this study. The major problem left here is the validity of the divergence obtained directly from the raw analysis of u and v, which is to be discussed in the following sub-sections.

3.7.2 January mean velocity potential field

Fig. 18 (upper panel) shows the mean χ-field at 200 mb during January 1979. A marked minimum is located at $15°S$, $175°E$, and two other areas of minimum are noted over South America and southern Africa, all three corresponding to the areas of climatologically active convection (and also corresponding to the anticyclonic circulations in flow field as shown in Fig. 8). Three maxima are observed in each hemisphere, located roughly over the surface high pressure areas (see Figs. 1 and 3).

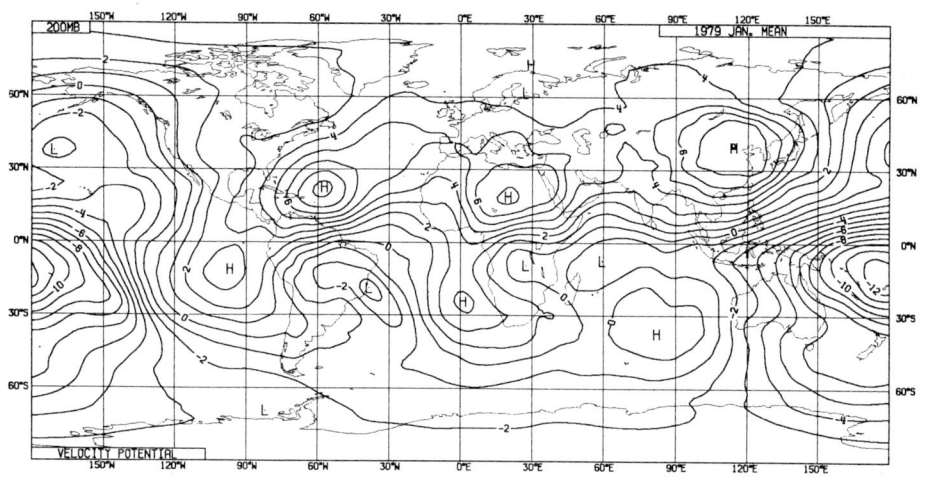
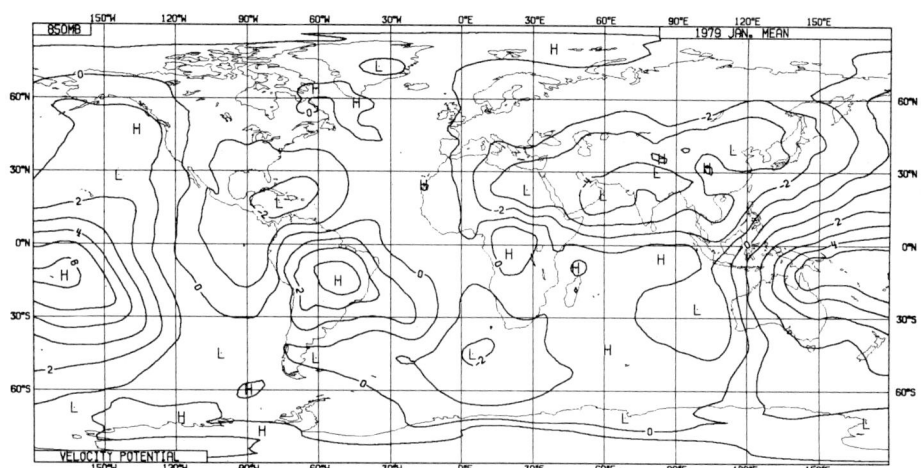

Fig. 18 January 1979 mean velocity potential field 200 mb (top) and 850 mb (bottom). Units in x 10^6 m^2 sec^{-1}. Contour interval is 10^6 m^2 sec^{-1}.

The divergent component of the wind is obtained from the relations

$$u_\chi = \frac{1}{a\cos\phi}\frac{\partial \chi}{\partial \lambda}$$
$$v_\chi = \frac{1}{a}\frac{\partial \chi}{\partial \phi}$$
(3)

the wind vector being directed from χ minimum to maximum. From this relation, we can roughly estimate the vertical circulations, i.e. upward over χ minimum and downward over χ maximum. The vertical circulations thus estimated in Fig. 18 are the three local Hadley circulations located at around 120-150°E, 30°E and 60°W, the strongest circulation being the first one with ascent over the south western Pacific and descent over Siberia. In the southern hemisphere, circulations in the east-west direction are more evident, all of them suggesting rising air over the continent and sinking air over the ocean. The velocity potential pattern at 850 mb (Fig. 18, lower panel) shows good negative correlations with 200 mb, supporting the existence of these vertical circulations. Considering that the analysis does not have any constraints on the vertical distributions of divergence except through the initial guess field, it is remarkable that the analysis scheme and the data (mostly cloud winds from satellites) can produce such a consistent velocity potential field.

3.7.3 Validity of the analysis

The velocity potential field during the northern hemisphere winter is presented by Krishnamurti et al (1973). Their computation is based on the analysis during December 1969 by the U.S. National Meteorological Center. The general pattern of the velocity potential agrees with the FGGE analysis but some of the details are quite different. This comparison is not conclusive since the annual and monthly variation of velocity potential is not well known and also the NMC analyses are based on poor data coverage in the tropics with no cloud motion vectors.

For better qualification of the analysis of the divergence field, some measure of convective activity is helpful. This is performed by Julian (1980) using archived satellite infrared digitized data at the National Center for Atmospheric Research. His study shows that the location of velocity potential minima and convective activity agrees very well in the monthly average and even in the daily chart. This fact provides a good basis for the validity of the divergence analysis, at least from a qualitative point of view.

3.7.4 Vertical distribution of the divergent part of wind

The distributions of velocity potential in the vertical have not been performed before simply because proper observations were not available especially in the lower to middle troposphere.

To exhibit the vertical distributions in a physical way, the kinetic energy of the divergent part of the wind is computed. Then, the divergent kinetic energy is separated into u-energy and v-energy, i.e.

$$K_{u\chi} = \frac{u_\chi^2}{2}$$
$$K_{v\chi} = \frac{v_\chi^2}{2}$$
(4)

$K_{u\chi}$ and $K_{v\chi}$ are the measure of the strength of the east-west and the local Hadley circulations, respectively. Furthermore, each kinetic energy is expressed in terms of zonal Fourier components to separate the space scales.

We present in Figs. 19 and 20 only two wave components (wavenumber 1 and 3), which have larger magnitude than other scales. In general, divergent kinetic energy is very small compared to the total kinetic energy (less than 5%). However, they are very important since all the conversion from potential (available) energy to kinetic energy takes place via the divergent part of the wind (Chen and Wiin-Nielsen, 1976). The large divergent kinetic energy is confined near the tropopause, in the boundary layer (below 850 mb) and in the northern hemisphere stratosphere.

An interesting character of the divergent kinetic energy is that the u-energy tends to have fairly large north-south scale, while the v-kinetic energy is concentrated in relatively narrow latitudinal bands (10° - 15°). This is very clear in the u-energy of zonal wave number 1, which has the scale of the distance between the two poles. (This is also confirmed in the subsequent February mean.)

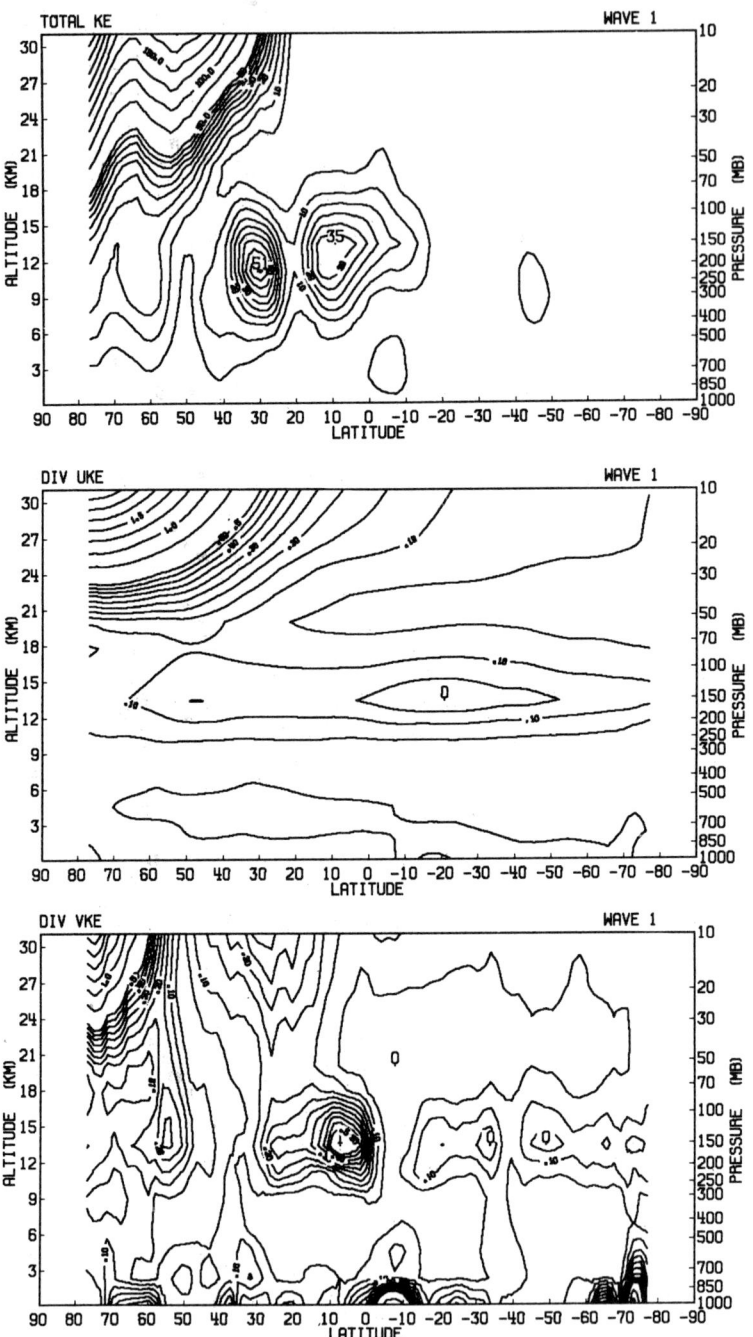

Fig.19 Meridional cross sections of the total kinetic energy (top) divergent u-kinetic energy (middle) and divergent v-kinetic energy (bottom) for zonal wavenumber 1, unit in $m^2\ sec^{-2}$. Contour interval is $5\ m^2\ sec^{-2}$ for the total kinetic energy ($25\ m^2\ sec^{-2}$ for values greater than $50\ m^2\ sec^{-2}$) $0.05\ m^2\ sec^{-2}$ for the divergent kinetic energies ($0.25\ m^2\ sec^{-2}$ for the values greater than $0.5\ m^2\ sec^{-2}$). The contours poleward of $80°N$ and $80°S$ are suppressed to prevent excessive crowding of the contours.

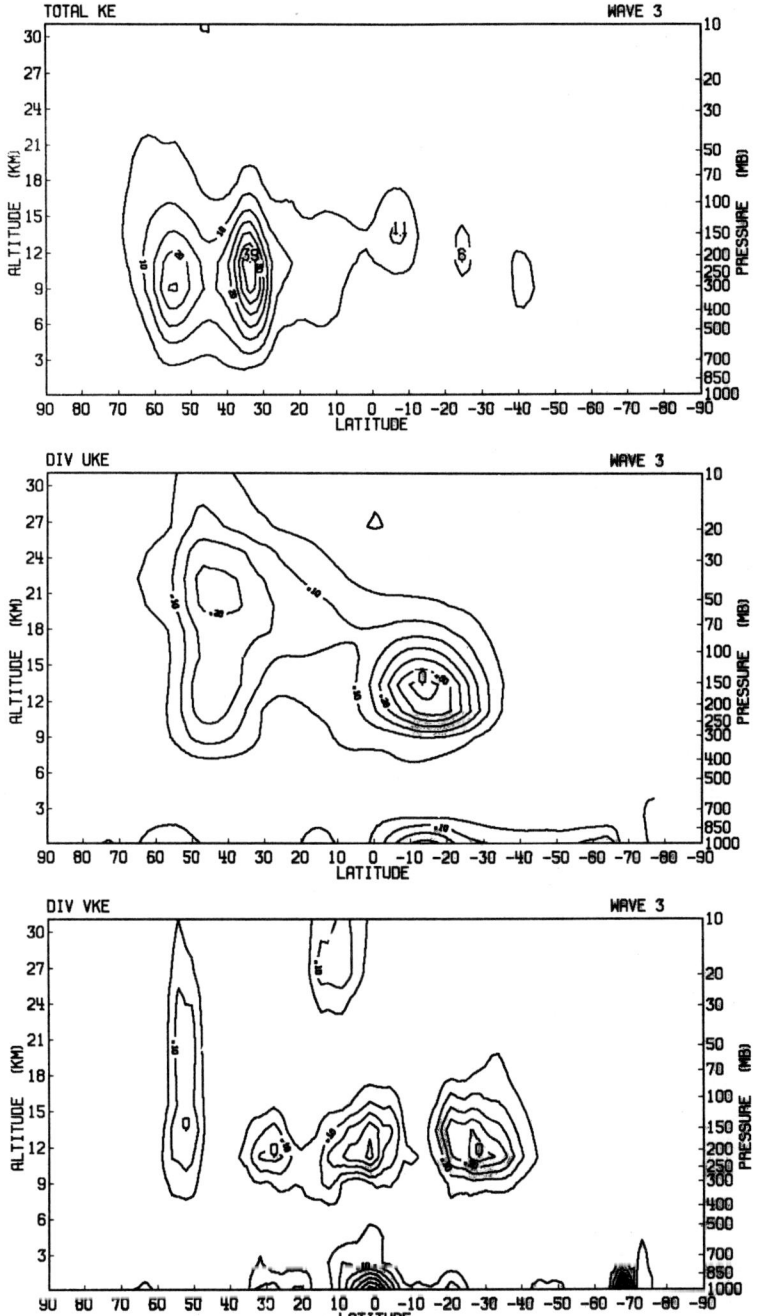

Fig.20 Same as Fig.19 but for zonal wavenumber 3.

There is some agreement in the location of the maximum of total kinetic energy and the divergent v-kinetic energy in wavenumber 1, at $10^O N$, suggesting conversion of potential to kinetic energy on this scale. Zonal wavenumber 3 indicates the existence of the east-west circulations at $10^O S$. The $K_{u\chi}$ maximum is situated at the middle of two maxima of $K_{v\chi}$ showing a very narrow zone of forcing at $10^O S$.

3.8 Evaluation of the analysis based on monthly averages

We have examined several January averaged fields of the ECMWF FGGE analysis against available climatologies in this section. Significant deviations from the 'old climatologies' are found in some of the fields. However, these deviations are within the annual variability or are explained by the improvement of the data coverage during FGGE, part of the latter conclusion being based on the examination of the daily analysis and the fit of the analysis to the observations during the monitoring of operational FGGE III-b data production. Furthermore, the velocity potential computed from raw analysis winds shows a very reasonable pattern which agrees quite well with convective activity as well as with climatological features. This physical consistency proves, among other things, that the analyses are of good quality.

4. KINETIC ENERGY BUDGET IN WAVENUMBER DOMAIN

In this section, we discuss kinetic energy budget calculations applied to the ECMWF analysis. Two major objectives should be noted. One is to examine further the physical consistency of the analyses, and the other to look into the energetics of the atmosphere in some detail.

4.1 Separation of scales

In formulating energy budget equations, there are several methods to separate the space scales of the motion field. Decomposition into the zonal averages and the deviations from them is practical and commonly applied (e.g. Oort, 1964) but it cannot reveal the different energetical nature of the very large scales and smaller scales which are known to be quite different. The space scale can be expressed either by the zonal Fourier coefficients at each latitude or by the spherical harmonic coefficients. The spherical harmonic analysis is very attractive from a mathematical point of view and also for describing the statistical character of the transient disturbances

(Baer, 1972). However, it seems to have some weakness when applied to the atmosphere, which has large inhomogeneity in the long time averages. Table 2 shows spherical harmonic spectra* of the mean FGGE January analysis at 200 mb.

---

* Only the rotational part of the winds are considered. The spherical harmonic component of the kinetic energy is expressed as

$$K_\ell^m = \frac{\ell(\ell+1)}{2a^2} (2-\delta_{om}) |\psi_\ell^m|^2 ,$$

where δ_{om} is a delta function, $\delta_{om} = 1$ for $m = o$ and $\delta_{om} = o$ for $m \neq o$, and ψ is a streamfunction defined as

$$\nabla^2 \psi = \frac{1}{a \cos\phi} \left\{ \frac{\partial v}{\partial \lambda} - \frac{\partial}{\partial \phi} (u \cos\phi) \right\}$$

It is noted that the contribution of the divergent part of the wind to the total kinetic energy is very small, as has been discussed in Sect. 3.7.4.

Table 2. Spherical harmonic kinetic energy spectra of the rotational part of the wind at 200 mb for the 1979 January mean analysis. The numbers in each column indicate energies for the given zonal wavenumber index (m) with increasing value of ℓ-m from bottom to top, ℓ being the order of the Legendre polynomial. Units are in m^2 sec^{-2}. Only the values for $1 \leq m \leq 7$ and $o \leq \ell$-m ≤ 7 are shown.

ℓ-m	m=1	m=2	m=3	m=4	m=5	m=6	m=7
7	1.1	0.9	1.3	0.3	0.0	0.0	0.0
6	7.9	0.1	1.8	1.0	0.0	0.0	0.1
5	3.8	0.4	1.4	0.5	0.2	0.0	0.1
4	0.2	3.3	2.1	2.4	0.4	0.1	0.1
3	10.1	6.9	6.3	2.6	0.8	0.1	0.4
2	1.3	0.5	0.3	0.4	0.7	0.6	0.6
1	0.3	0.7	1.1	0.5	1.8	1.9	0.6
0	0.0	0.1	0.1	0.2	0.0	0.2	0.2

The energy spectra spreads in many spherical harmonic components
and it is very difficult to see systematic distributions in zonal
wavenumber index (m) or in the degree of the Legendre polynomial (ℓ).
There is a fair indication that the kinetic energy peaks appear
in the components of odd values of ℓ-m showing dominance of anti
symmetric modes(with respect to the equator). This broad spectra
of the monthly means discourages us from using spherical harmonic
decomposition. Hence, we have decided to use zonal Fourier
decomposition, and tried to examine the latitudinal variations.
The zonal Fourier decomposition has an advantage that the dominant
mode can be easily identified in the original field, the disadvantage being the variation of actual wave lengths with latitude for
the same wave number.

4.2 The energy budget equations in wavenumber domain and the computational method

The equations of energetics in wavenumber domain was first derived
by Saltzman (1957). The temporal change of the wave components
of kinetic energy at any latitude are written as follows.

$$\frac{\partial K(o)}{\partial t} = \sum_{m=1}^{\infty} M(m) + C(o) + F(o) + W(o) + D(o)$$

$$\frac{\partial K(m)}{\partial t} = - M(m) + L(m) + C(m) + F(m) + W(m) + D(m)$$

(5)

In the above equations, K(m) is the kinetic energy of wavenumber m,
m = o being the zonal average, M is the energy exchange between the zonal
motion and the waves, C is the conversion from available potential
energy, L is the kinetic energy exchanges between the waves, F is
the boundary flux convergence both in the horizontal and in the
vertical, W is the pressure work at the boundaries and D is
dissipation. The complete forms of the energy conversion terms
on the right hand side of Eq.(5) are described in the Appendix.

The energy conversion terms are evaluated in the box surrounded by
the grid points between the two standard pressure levels (Fig. 21).
As can be seen in the Appendix, the conversion terms consist of
several pairs of terms related by the continuity equation. Finite
differences in the north-south and in the vertical require some caution
to avoid any fictitious generation of mass and energy, which causes
serious errors in the evaluation of some of the terms. The grid

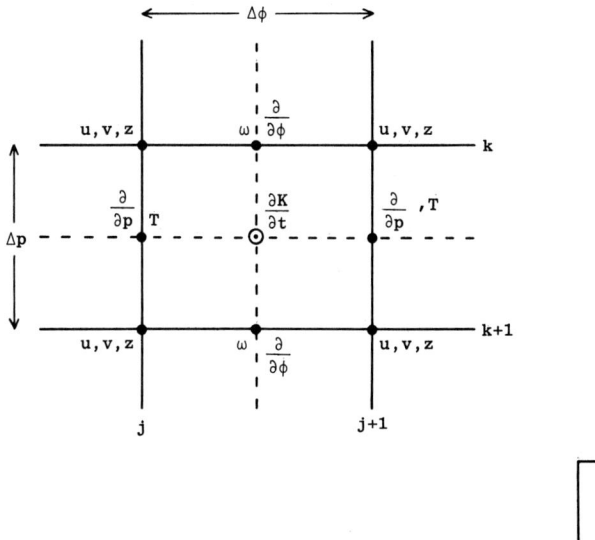

Fig.21 Configuration of grids used in energy budget calculations. j denotes index for the north-south grids and k for the pressure levels.

system (Fig. 21) is chosen for this reason.

The temperature is computed from the thickness between the two pressure levels. No vertical interpolation is involved. The vertical velocity (ω) is computed from analyzed (uninitialized) u and v assuming ω = o at 1000 mb. No constraint on ω is applied in the calculation. Moreover, no particular care is taken over the high topography area (u,v and z are analyzed even if the pressure surface is below the ground). The upper limit of the wavenumber is set to 30 in all the calculations.

The dissipation terms are computed as residuals knowing the time tendency of the kinetic energy from the analyses 12 hour apart. This residual was found to be as large as other terms even in the monthly average, with both negative and positive signs. This suggests uncertainties in evaluating some of the conversion terms due to time and space truncation errors (including the effect of waves shorter than wavenumber 30), and to physical inconsistencies of the analysis. To find the terms responsible for these uncertainties in the evaluations of the conversion terms, 12 hourly values of each term are mapped and checked particularly when the residuals become large positive. It was found, in most cases, that the terms related to the divergence are the source of the problem. This is in fact expected from a somewhat crude way of computing vertical motion as noted earlier. This indicates that although a long time averaged large scale divergence field seems to show very reasonable distributions as discussed before (3.7.2), individual analyses are not accurate enough in magnitude and in the correlation with other variables to give reasonable estimates of the energy conversions. To overcome these problems, use of initialized vertical motion is very attractive but still not adequate in low latitudes (see 3.7.1). Thus, since it is not so simple to obtain accurate vertical motion fields, we have decided to eliminate all the terms related to the divergence (terms with ω as well as the pressure work term at the boundaries which relates to the divergent part of the wind) and included them in the residual term. This implies that in Eqn. 5, part of the terms, M,L and F (terms related to the vertical advections), and the terms C, W are combined with the term D (see Appendix).

4.3 Kinetic energy budget during January 1979

In order to examine the energetics in the tropics as well as in high latitudes, we have selected the layer between 250 mb and 200 mb, where tropical activity is greatest (see Figs. 12-14, 19-20). Fig. 22 shows energy diagrams for the northern hemisphere ($90°N$ - $20°N$), tropics ($20°N$ - $20°S$) and for the southern hemisphere ($20°S$ - $90°S$).

4.3.1 Northern hemisphere (Fig. 22, left)

In the northern hemisphere, zonal-wave interactions (M) are smaller than the wave-wave interactions (L). The zonal motion receives kinetic energy from waves as a total (some waves receive energy from the zonal but not significantly). The wave-wave interaction shows a large output of energy from baroclinic waves (wavenumber, $m = 7$ to 10) to long (particularly wavenumber 1) and short waves (with some exceptions). The residual terms indicate input of energy to baroclinic waves suggesting conversion from potential energy to kinetic energy on this scale. A large output of kinetic energy in wavenumbers 1 and 2 is noted. The overall energy exchanges are in very good agreement (although details are not quite the same) with similar studies performed by various people, which are summarized by Saltzman (1970).

4.3.2 Southern hemisphere (Fig. 22, right)

The most striking difference from the northern hemisphere is that the zonal-wave interaction is larger than the wave-wave interactions. This is physically quite convincing since the ultralong waves in the southern hemisphere are not as active as in the northern hemisphere due to the weaker forcing by surface irregularities. This also suggests the importance of surface conditions in generating and maintaining the ultralong waves.

Zonal wave interactions are much simpler than in the northern hemisphere. All the waves supply energy to the zonal motion, peaking at $m = 5$ (the dominance of which has been noted in Fig. 4). Wave-wave exchanges are not so simple but in general, conversions from baroclinic waves ($m = 7$ to 10) to other waves are taking place.

ENERGY DIAGRAM (GLOBAL) JANUARY 1979   250 mb—200 mb

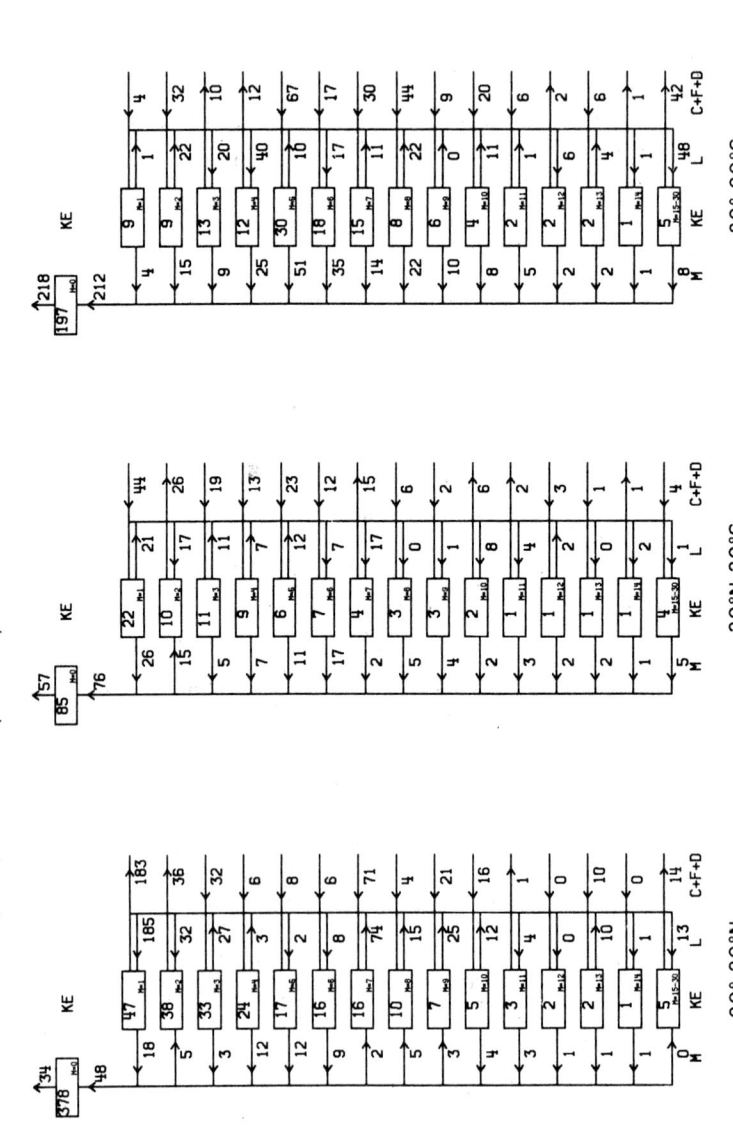

Fig. 22   Energy diagram for the northern hemisphere (left) tropics (centre) and southern hemisphere (right) in the layer between 200 and 250 mb during January 1979 obtained from ECMWF analyses. The large numbers in the boxes are the area averaged kinetic energies of the zonal Fourier component m indicated by the small number. The box at the top of the diagram is for the zonal mean field (m=o). The arrows and lines at the left side of the diagram connecting the boxes with the one at the top indicate zonal-wave interactions (M) and their directions. Similarly the lines connecting the boxes on the right side of the diagram indicate wave-wave exchanges (L). The other lines on the right and at the top indicate the residuals (the sum of conversion from potential energy, boundary fluxes, dissipations, etc). Units are in $m^2\ sec^{-2}$ for energies, $m^2\ sec^{-3}$ for the conversions.

4.3.3 Tropics (Fig. 22, centre)

The tropical energetics are quite different from those in high latitudes. The energy spectrum is simpler, having its maximum at m=1 and decreasing almost monotonically as the scale decreases (the number in the boxes of Fig. 23 is the energy of each wave in $m^2 sec^{-2}$ unit). This is considered to be due to the weak baroclinic wave activity in the tropics. The zonal motion (weak westerlies, ~ 13 m/s) is maintained by the waves (except by m = 2). The sources of kinetic energy for smaller scales (m ⩾ 6) are the ultralong waves (m = 1, 3 to 5) which are maintained by C+F+D, most likely by the conversion from potential energy suggested in Figs. 16, 20 and 21. The energy conversion to medium scale waves (m = 7 to 11) suggests development of waves by barotropic instability.

In summary, the most important differences of the energetics in the three regions selected above can be summarised as follows:

i) The major kinetic energy source in high latitudes is in the baroclinic waves

ii) In the northern hemisphere, this energy source is used mainly to maintain ultralong waves

iii) In the southern hemisphere, it is used to maintain the zonal motion

iv) The major kinetic energy source in the tropics is the ultralong waves forced by the organised tropical convection (and probably by the middle latitudes).

It is noted that the tropical energetics obtained here do not quite agree with those of Krishnamurti et al (1973) in the conversion between the zonal flow and the ultralong waves (m = 1 to 3 receive energy from the zonal flow in their study). We have repeated the calculations by taking the same domain as Krishnamurti's ($15°N - 15°S$). In this domain, m = 1 still provides energy to the zonal flow but less in magnitude (15 units vs 26 units). The m - 3 and zonal conversion reversed in sign (m = 3 receives 5 units). Since the differences between the two studies can be due to various other reasons (month to month, year to year change, data coverage etc.), we have not tried to examine this discrepancy further.

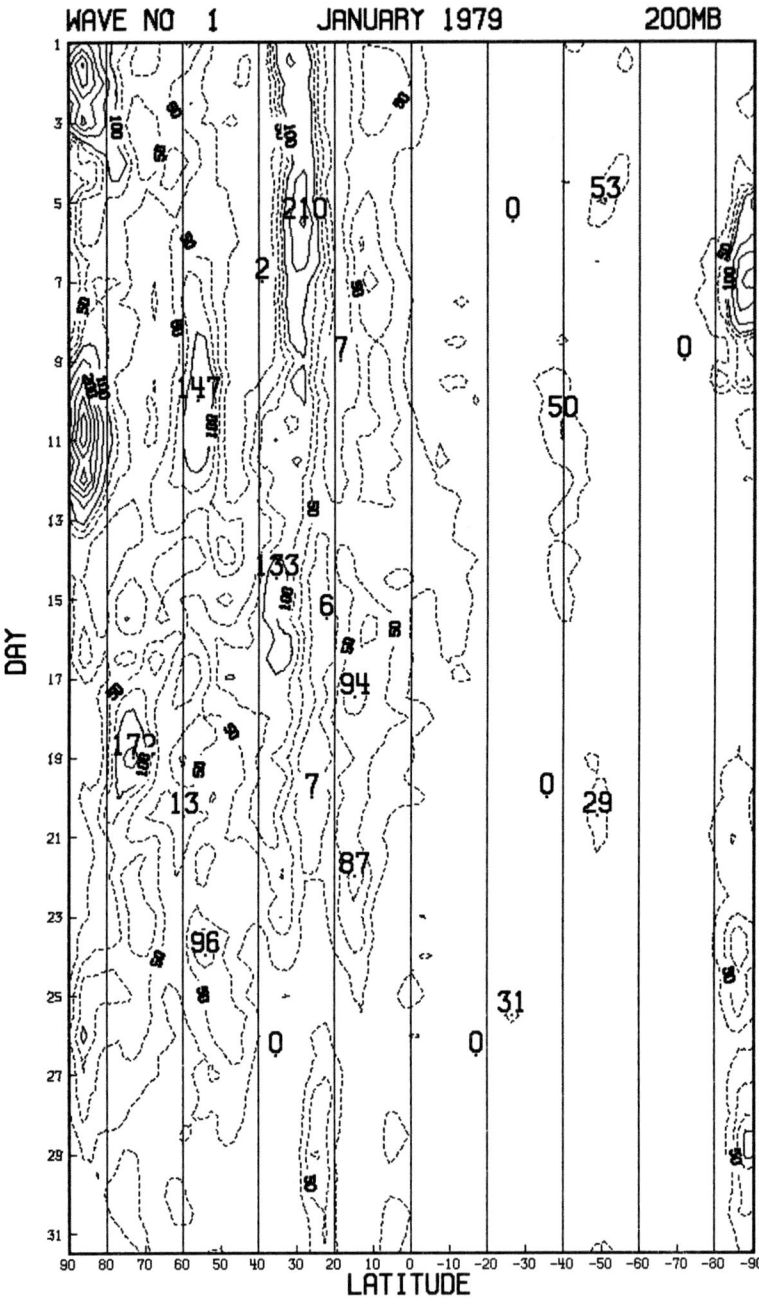

Fig.23 Latitude (abscissa) and time (ordinate) variations of the kinetic energy of zonal wavenumber 1. Unit in $m^2 sec^{-2}$. Contour interval, 25 $m^2 sec^{-2}$.

4.4 Latitudinal variation of the energetics

There are two major reasons to examine latitudinal variations of the energetics. One is to understand the somewhat complex nature of the ultralong waves in the northern hemisphere (Fig. 22) and the other is to look into the cause of the differences in detail of the energetics among various studies, some of which are noted in the previous section.

Fig. 23 shows latitude-time variations of the kinetic energy of $m = 1$ at 200 mb. The most apparent character of the wave is the very narrow north-south scales (about $10°$ from minimum to minimum) with maxima at around 10, 30 and $60°N$. Tropical wavenumber 1 has a somewhat larger scale. The small N-S scale agrees with the synoptic observations of the narrowness of the local westerly jet. Similar figures for other waves (not shown) indicate that $m = 2$ also has narrow N-S scale, but as m increases, there is a tendency that the N-S scale increases. For the baroclinic waves (m=7-10) the N-S scales are $20-30°$ latitude from minimum to minimum. The small N-S scales of the ultralong waves suggest large latitudinal variations of the energetics in this scale.

Another noteworthy feature of the ultralong waves are their temporal variations. The kinetic energy of $m = 1$ at $30°N$, for example, goes through considerable variations from a very intense state (Jan 1-7) to being almost non existent (Jan 23-27). Similar large variations (not so clearly coherent with other waves) are observed in other ultralong waves. The variations of the baroclinic waves ($m = 7-10$) are more intermittent with a period of 4-5 days (not shown). This analysis indicates the importance of examining temporal variations of the energetics.

To examine the latitudinal variation of the energetics, monthly averaged values of M,L,F and D for each wavenumber are plotted against latitudes. The results for the zonal mean and for the wavenumber 1 are shown in Fig. 24.

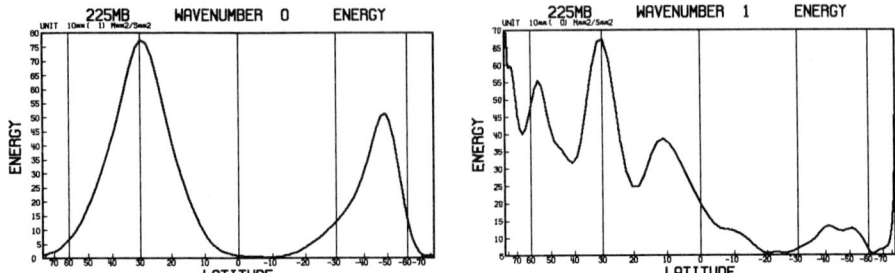

Fig.24a Latitudinal variations of the energy for various wavenumbers. Note that the ordinate scales are not the same for all the waves. The latitude scales are weighted by the area.

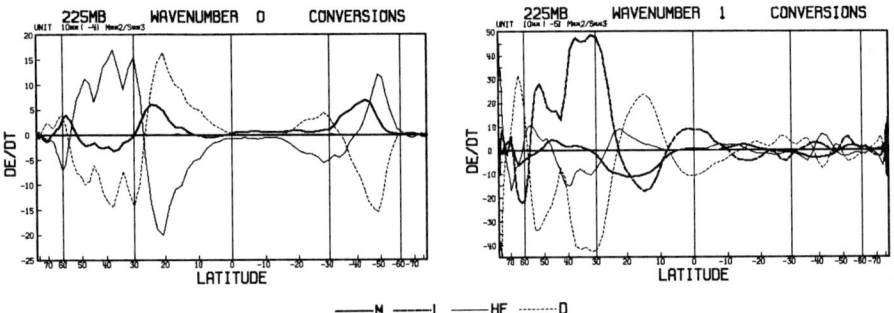

Fig.24b Same as Fig. 24a but for the energy conversions. Thick solid, thick dashed, thin solid and thin dashed lines indicate zonal-wave, wave-wave, flux and residual terms, respectively.

The zonal conversion (upper left figure) shows that the largest contribution to the zonal motion is the flux convergence term, which is negatively correlated with the smaller barotropic conversion term*. The distributions of these two terms indicate that the westerly momentum is generally transported northward in the northern hemisphere middle latitudes (southward in the southern hemisphere). The signs of M and F depend on the gradient of the zonal mean westerlies. The positive and the negative residual to the south and the north of the westerly jet indicates generation and destruction of the kinetic energy by the cross isobaric meridional circulation of the tropical Hadley and the middle latitude Ferrel cells, respectively.

The energetics of the wave show considerable variations with latitude. By the careful examination of the latitudinal variation of energetics, we are able to divide the northern hemisphere into three zones, which indicate quite different energetics amongst them.

4.4.1 Polar region ($90°N - 70°N$. Fig. 25, left)

The energetics in this zone are very much different from the other zones. The zonal motion provides a large amount of energy to m = 1-6, indicating the barotropically unstable nature of the waves. Energy exchanges between the waves show a source of kinetic energy in m = 2 and 4, all other waves receiving energy from them. Since the actual wavelength for the given wavenumber is much smaller than that in the lower latitudes (about 14000 km for m = 1 at $70°N$), it may not be proper to use the word 'ultralong waves' in this zone. It is also difficult to classify baroclinic waves from this point of view. The relation with the evolutions of the stratospheric polar vortex seems to be essential in this region, but this is beyond the scope

---

*
It may be misleading to call the term 'barotropic conversion', because part of the flux convergence term contains wave-zonal interaction at the boundaries as shown in Appendix. The term M is more properly named as the transport of the westerly momentum up or down the gradient of the zonal westerlies. Similar terminology applies to L and F.

ENERGY DIAGRAM (NORTHERN HEMISPHERE) JANUARY 1979   250 mb–200 mb

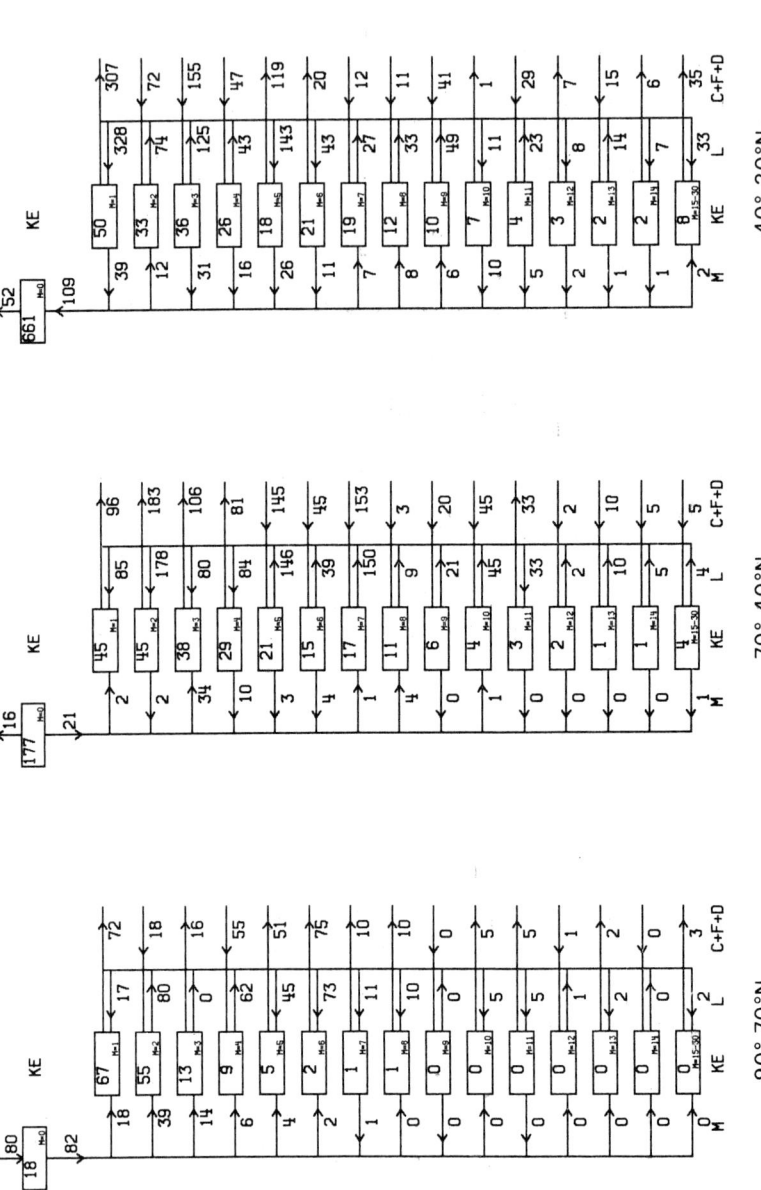

Fig.25 Same as Fig.22 but for the high latitude zones (left) middle latitude zones (centre) and for substropical zones (left)

of this study. The use of the zonal Fourier analysis may not be a proper method to decompose the scales in these high latitudes. The use of spherical harmonics may be more appropriate.

4.4.2 Middle latitude zone ($70°N - 40°N$, Fig. 25, centre)

In this region, the zonal motion still provides its energy to the waves, mainly to $m = 3$. The role of wavenumber 3 as a drain on the energy of the zonal motion is noted by Wiin-Nielsen et al (1963) and Yang (1967). Conversions from all other waves are not significant (except $m = 4$ providing energy to the zonal). On the other hand, the wave-wave exchanges are significant. The kinetic energy sources exist in the baroclinic waves ($m = 5-10$), providing their energy to the ultralong waves ($m = 1-4$). Down the scale cascade of the energy is small. The residual terms of the ultralong waves are consistently negative. These energy exchange processes strongly indicate that the ultralong waves are dynamically forced (contrasted to thermally-forced) in these latitudes.

4.4.3 Subtropical zone ($40°N - 20°N$. Fig. 25, right)

This is the zone where the subtropical jet has its maximum intensity ($30°N$). The zonal to wave interaction reverses its sign from those in the other two areas, the zonal motion being maintained by the supply of kinetic energy from the waves. The wave-wave interactions are quite different from those in the middle latitude zone. The kinetic energy source exists in two scales, i.e. the baroclinic waves ($m = 7-11$, except 10), and the ultralong waves ($m = 2-4$) except wavenumber 1. A large amount of energy is transferred to wavenumber 1, which is consumed by the zonal-wave interaction, conversion to potential energy, flux across the boundaries, dissipation, etc. These conversions indicate that the forcings in the scale of $m = 2-4$, possibly by the heating and by the boundaries are essential in this area. It is noted that the major part of the Himalayas and some part of the Rockies are included in this zone, which may be related to the forcing in these scales. The strong sink of kinetic energy in wavenumber 1 is quite interesting. It indicates that this wave is dynamically forced through the interactions between the waves $m = 2,3$ and 4.

5. SUMMARY AND CONCLUSIONS

The January 1979 analyses produced by the 4-dimensional assimilation scheme at the ECMWF using complete FGGE observations (II-b data sets) have been examined through climatological and kinetic energy budget studies.

The comparisons with the long time averaged January (and seasonal) climatologies indicate the following:

   i) The northern hemisphere is somewhat anomalous due to the strong blockings persisting during the month.

   ii) Significant differences in the flow fields over the oceans indicate the impact of the FGGE observations. The time variabilities of the motion field over the tropics in the upper troposphere are somewhat smaller than the climatological means. The momentum transports by the transient eddies are much smaller over a similar area and levels.

   iii) In the southern hemisphere, surface highs and lows are more intense. Upper level flow patterns have more intense troughs and ridges, again indicating the impact of the FGGE observations. Momentum transports by the standing eddies are larger and the pattern different.

The analyses of the tropical divergence fields have been examined by mapping the velocity potential field. From the comparison with the initial guess field, a large part of the analysed divergence has been found to be produced by the observations. The initialization procedure (non linear normal mode initialization) has a weakness of reducing the analysed divergence in the tropics.

The large scale velocity potential fields depict three local Hadley circulations between the areas of active tropical convection and the northern hemisphere continents, and the east-west circulations between the continents and the oceans in the southern hemisphere. The circulations have largest kinetic energy in the planetary boundary layer and at the tropopause.

A kinetic energy budget in the zonal wavenumber domain is performed on the analyses. The vertical motion field directly computed from uninitialized winds causes undesirably large residuals in computing the energy conversion terms. Other means of computing vertical motion must be investigated. The current computation excludes all the divergence related terms and includes them in the residual term.

The kinetic energy budget in the northern hemisphere agrees very well with the results compiled by Saltzman (1970), showing the role of baroclinic waves as kinetic energy sources for other scales of motion. The wave-wave interactions are stronger than the zonal-wave interactions, also in agreement with Saltzman. Some differences are observed in the exchanges of each individual wave. In the southern hemisphere, the wave-zonal interaction dominates the wave-wave interactions indicating the less important role of the ultralong waves, making a clear contrast with the northern hemisphere energetics. The energetics in the tropics are quite different from those in high latitudes. The kinetic energy source is in the very large scale and the medium scale disturbances are maintained by receiving energy from very large scales. This result is similar to the energetics in the tropics during the northern hemisphere summer.

The latitude-time variation of the kinetic energy of the ultralong waves shows narrow north-south scale (about $10°$ latitude from minimum to minimum) with large temporal variations. As a result, energetics of the ultralong waves change significantly with latitude and time. The northern hemisphere is shown to be divided into several zones according to the different monthly averaged energetical nature of the motion field. Remarkable differences of the properties of the ultralong waves are found between the middle latitude zone ($70°N - 40°N$) and the subtropical zone ($40°N - 20°N$). In the middle latitude zone, all the ultralong waves are maintained by the baroclinic waves through the nonlinear interactions, while in the subtropical zone, ultralong waves except wavenumber 1, are also the energy sources for other waves. These energetics suggest that, in the latter zone, where nearly the entire Himalayas and part of the Rockies are included, the ultralong waves are externally (most likely thermally) forced and in the former zone, they are dynamically forced by the interactions with the baroclinic waves. The wavenumber 1 in the subtropical zone has considerable time variability and the transient parts are largely controlled by the nonlinear interactions among the waves.

The various climatological and energetic computations performed in this paper clearly reveal that the analyses of the FGGE data by the ECMWF 4-dimensional assimilation scheme are of good quality and suitable for diagnostic studies. The quantitative evaluation of the vertical motion field, however, still seems to have some difficulties. Use of the initialized field or other means have to be investigated depending on the purpose.

Acknowledgements

The author gratefully acknowledges his colleagues of the FGGE group at the ECMWF for their help in performing this study.
Drs. A. Simmons and D. Shaw have been very helpful in improving the manuscript. Thanks are also due to Drs. L. Bengtsson, A. Hollingsworth and P. Julian for their many helpful discussions and comments.

APPENDIX

The conversion terms of the kinetic energy equation in the zonal wavenumber domain (Fig. 5) are expressed as follows.

(i) Zonal-wave interaction term (M)

$$M(m) = M_1(m) + M_2(m) + M_3(m)$$

$$M_1(m) = \frac{2}{a} \left\{ \frac{\partial U(o)}{\partial \phi} \text{Re}\,[U(m)\,V(-m)] + \frac{\partial V(o)}{\partial \phi} \text{Re}\,[V(m)\,V(-m)] \right\}$$

$$M_2(m) = \frac{2\tan\phi}{a} \left\{ U(o)\text{Re}[U(m)\,V(-m)] - V(o)\text{Re}[U(m)\,U(-m)] \right\}$$

$$M_3(m) = 2 \left\{ \frac{\partial U(o)}{\partial p} \text{Re}\,[U(m)\,W(-m)] + \frac{\partial V(o)}{\partial p} \text{Re}\,[V(m)\,W(-m)] \right\}$$

(ii) Wave-wave interaction term (L)

$$L(m) = L_1(m) + L_2(m) + L_3(m)$$

$$L_1(m) = -\sum_{k=1}^{\infty} \frac{2m}{a\cos\phi} \text{Im}\,[\Psi(U,U,U) + \Psi(V,V,U)]$$

$$\qquad\qquad - \frac{2}{a} \text{Re}\,[\Psi(\frac{\partial U}{\partial \phi},U,V) + \Psi(\frac{\partial V}{\partial \phi},V,V)]$$

$$L_2(m) = -\sum_{k=1}^{\infty} \frac{2\tan\phi}{a} \text{Re}\,[-\Psi(U,V,U) + \Psi(V,U,U)]$$

$$L_3(m) = \sum_{k=1}^{\infty} 2\,\text{Re}\,[\Psi(\frac{\partial U}{\partial p},U,W) + \Psi(\frac{\partial V}{\partial p},V,W)]$$

(iii) Baroclinic conversion term (C)

$$C(o) = -\frac{R}{p} W(o)''\,T(o)''$$

$$C(m) = -\frac{2R}{p} \text{Re}\,[W(m)\,T(m)]$$

(iv) Flux term (F)

$$F(m) = F_1(m) + F_2(m)$$

$$F_1(o) = -\frac{1}{a\cos\phi}\frac{\partial}{\partial\phi}\cos\phi\left\{V(o)K(o) + 2U(o)\sum_{k=1}^{\infty}\text{Re}\,[U(k)\,V(-k)]\right.$$

$$\left. + 2V(o)\sum_{k=1}^{\infty}\text{Re}\,[V(k)\,V(-k)]\right\}$$

$$F_2(o) = -\frac{\partial}{\partial p}\left\{W(o)K(o) + 2U(o)\sum_{k=1}^{\infty}\text{Re}\,[U(k)\,W(-k)]\right.$$

$$\left. + 2V(o)\sum_{k=1}^{\infty}\text{Re}\,[V(k)\,W(-k)]\right\}$$

$$F_1(m) = -\frac{1}{a\cos\phi}\frac{\partial}{\partial\phi}\,2\cos\phi\left\{\sum_{k=1}^{\infty}\text{Re}\,[\Psi(U,U,V) + \Psi(V,V,V)]\right\}$$

$$F_2(m) = -\frac{\partial}{\partial p}\,2\sum_{k=1}^{\infty}\text{Re}\,[\Psi(U,U,W) + \Psi(V,V,W)]$$

(v) Pressure work term (W)

$$W(m) = W_1(m) + W_2(m)$$

$$W_1(o) = -\frac{g}{a\cos\phi}\frac{\partial}{\partial\phi}\cos\phi\left\{V(o)\,Z(o)''\right\}$$

$$W_2(o) = -g\frac{\partial}{\partial p}\left\{W(o)\,Z(o)''\right\}$$

$$W_1(m) = \frac{2g}{a\cos\phi}\frac{\partial}{\partial\phi}\cos\phi\,\text{Re}\,[V(m)\,Z(-m)]$$

$$W_2(m) = -2g\frac{\partial}{\partial p}\,\text{Re}\,[W(m)\,Z(-m)]$$

(vi) Dissipation term (D)

$$D(o) = U(o)\,X(o) + V(o)\,Y(o)$$

$$D(m) = 2\,\text{Re}\,[U(m)\,X(-m) + V(m)\,Y(-m)]$$

(vii) Kinetic energy (K)

$$K(o) = \frac{1}{2} \{U(o)^2 + V(o)^2\}$$

$$K(m) = |U(m)|^2 + |V(m)|^2$$

In the above expressions, Ψ is an operator denoting,

$$\Psi(A,B,C) = A(m) \{B(-k) C(-m+k) + B(k) C(-m-k)\}$$

Im and Re denote imaginary and real part respectively.
The meaning of the symbols used are as follows

    $U(m)$: zonal Fourier component of u
    $V(m)$: zonal Fourier component of v
    $W(m)$: zonal Fourier component of ω
    $X(m)$: zonal Fourier component of the east-west component of the frictional force per unit mass.
    $Y(m)$: zonal Fourier component of the north-south component of the frictional force per unit mass
    $Z(m)$: zonal Fourier component of geopotential height.

    R : gas constant of dry air
    g : acceleration of gravity
    p : pressure
    ω : dp/dt

The double prime denotes the deviation from a north-south average:

$$\frac{1}{(\sin\phi_1 - \sin\phi_2)} \int_{\phi_1}^{\phi_2} A \cos\phi \, d\phi$$

The values in the diagrams (Figs. 23 and 26) are obtained by applying the above averaging procedures.

The divergence related terms are M_3, L_3, C, F_2 and W, which are set to zero in most of the computations, except in the case of the stationary part of the motion field.

References

Baer, F., 1972: An alternative scale representation of atmospheric energy spectra. J.Atmos. Sci., 29, 649-664.

Bengtsson, L. and P. Kallberg, 1980: Numerical simulation - Assessment of FGGE data with regard to their assimilation in a global data set. Proceedings of the International Conference on Preliminary FGGE data analysis and results. Bergen, Norway (in press).

Chen, T.C. and A. Wiin-Nielsen, 1976: On the kinetic energy of the divergent and nondivergent flow in the atmosphere. Tellus, 28, 486-497.

Julian, P.R., 1980: Data assimilation for the FGGE tropical observing system. Proceedings of ECMWF Seminar 1980 on Data Assimilation Methods, 375-398.

Krishnamurti, T.N., 1971: Tropical east-west circulations during the northern summer. J.Atmos.Sci., 28, 1342-1347.

Krishnamurti, T.N., M. Kanamitsu, W.J. Koss and J.D. Lee, 1973: Tropical east-west circulations during the northern winter. J.Atmos.Sci., 30, 780-787.

Lorenc, A., I. Rutherford and G. Larsen, 1977: The ECMWF analysis and data assimilation scheme - Analysis of mass and wind fields. ECMWF Technical Report No. 6, pp 47.

Lorenc, A.C., 1980: A global three-dimensional multivariate statistical interpolation scheme. Proceedings of the International Conference on Preliminary FGGE Data Analysis and Results. Bergen, Norway (in press).

Newell, R.E., J.W. Kidson, D.G. Vincent and G.J. Boer, 1972: The General Circulation of the Tropical Atmosphere and Interactions with Extratropical Latitudes. Vol. 1. The MIT Press, Cambridge, Mass., USA and London, UK, pp 258.

Oort,A.H., 1964: On estimates of the atmospheric energy cycle. Mon.Wea.Rev., 92, 483-494.

Saltzman, B., 1957: Equations governing the energetics of the large scales of atmospheric turbulence in the domain of wave number. J. Meteor., 14, 425-431.

Saltzman, B., 1970: Large-scale atmospheric energetics in the wavenumber domain. Rev.Geophys., 8, 289-302.

Temperton, C. and D.L. Williamson, 1979: Normal mode initialization for a multi-level gridpoint model. ECMWF Technical Report No. 11, pp 91.

Wiin-Nielsen, A., J.A. Brown and M. Drake, 1963: On atmospheric energy conversions between the zonal flow and the eddies. Tellus, 15, 261-279.

Yang, C.H., 1967: Nonlinear aspects of the large scale motion in the atmosphere. Sci. Report No. 08759-1-T, Dept. of Meteorology and Oceanography, University of Michigan, Ann Arbor.

APPENDIX

PROVISIONAL REPORT ON CALCULATION OF SPATIAL COVARIANCE AND AUTOCORRELATION OF THE PRESSURE FIELD[*]

Arnt Eliassen

Videnskaps-Akademiets Institutt for
Vaer - Og Klimaforskning
Oslo, Norway

1. DEFINITIONS

Let $z(x,y)$ denote the horizontal field of height of an isobaric surface, and let a bar designate average with respect to time. The deviation of the height from its mean value is

$$\zeta(x,y) = z(x,y) - \bar{z}(x,y) \tag{1}$$

The variance of z is

$$N(x,y) = \overline{\zeta(x,y)^2} \tag{2}$$

and the covariance between the heights in the points x_o, y_o and x,y

$$N(x,y,x_o,y_o) = \overline{\zeta(x,y)\,\zeta(x_o,y_o)} \quad . \tag{3}$$

The autocorrelation function is

$$R(x,y,x_o,y_o) = \frac{M(x,y,x_o,u_o)}{\sqrt{N(x,y)\,N(x_o,y_o)}} \quad .$$

These statistical characteristics of the pressure field are easily calculated from aerological data.

---

[*] Report reprinted from Rapport nr. 5,
Videnskaps-Akademiets Institutt for Vaer -
Og Klimaforskning, Oslo, Norway,
September 1954.

2. PROCEDURE AND RESULTS

The values of $\bar{z}$, N, M and R have been calculated for 15 stations in north-western Europe for the isobaric surfaces 1000, 700, 500 and 300 mb. The time averaging was taken over the months of December and January for the six years December 1948 – January 1954. The calculations were considered as tentative, and in order to save work, only the afternoon observations every fourth day were used, beginning December 2nd. Thus the mean values have been computed from 96 individual observations at each station; in some cases where observations were missing, the number is somewhat smaller. It should be noted in this connection that the additional information which could have been gained by using data more closely spaced in time would not have been proportional to the additional work involved, owing to the serial correlation of pressure.

The results of the calculations are shown in the tables below.

Table 1 shows the values of $\bar{z}$ at the four levels, in meters.

Table 2 shows the distribution of the variance N. As one would expect, there is a general increase of N with height; this increase is strongest in the southern part of the region considered, over the British Isles, and much weaker for the northernmost stations Jan Mayen and Skattöra.

Table 3 shows the covariance M relative to the station Liverpool. The maximum values are found a few hundred kilometers north of Liverpool.

Table 4 shows the autocorrelation function R relative to Liverpool. In the vicinity of Liverpool, the pattern is fairly symmetric around this station, but at greater distances there is a considerable distortion, with a more rapid decrease of R toward the west and a slower decrease toward the north-east. This distortion increases with height; thus R decreases with height at the weather ship J west of the British Isles, and increases with height at Jan Mayen. None of the stations considered are far enough from Liverpool to have negative values of R.

Table 1. Values of $\bar{Z}$ in meters, and number of observations (in parenthesis)

Station	1000 mb	700 mb	500 mb	300 mb
Liverpool	107 (96)	2946 (96)	5476 (96)	9008 (96)
Camborne	134 (96)	2990 (96)	5545 (96)	9102 (96)
Larkhill	128 (96)	2972 (96)	5514 (96)	9056 (96)
Hemsby/Downham M.	116 (96)	2954 (96)	5487 (96)	9016 (96)
Valentia	113 (96)	2974 (96)	5524 (96)	9084 (96)
Aldergrove	99 (96)	2938 (96)	5470 (96)	9003 (96)
Leuchars	79 (96)	2909 (96)	5429 (96)	8947 (96)
Stornoway	56 (96)	2883 (96)	5394 (96)	8898 (96)
Lerwick	50 (96)	2860 (96)	5362 (96)	8848 (96)
Weather-ship J 52.5°N, 20°W	77 (96)	2932 (96)	5478 (96)	9034 (96)
Weather-ship M 66°N, 2°E	13 (96)	2803 (96)	5267 (96)	8713 (91)
Jan Mayen	11 (96)	2750 (96)	5191 (96)	8596 (87)
Skattöra	25 (96)	2773 (96)	5213 (96)	8593 (64)
Gardermoen	83 (96)	2867 (96)	5342 (96)	8787 (85)
Sola	80 (96)	2892 (96)	5388 (96)	8877 (89)

Table 2. Variance (N), in dekameters^2

Station	1000 mb	700 mb	500 mb	300 mb
Liverpool	96	139	259	465
Camborne	81	122	213	401
Larkhill	78	121	240	437
Hemsby/Downham M.	89	131	265	460
Valentia	116	127	278	398
Aldergrove	111	150	263	469
Leuchars	139	166	293	507
Stornoway	156	185	312	573
Lerwick	170	192	330	533
Weather-ship J	163	189	280	421
Weather-ship M	177	181	270	465
Jan Mayen	150	135	198	318
Skattöra	153	171	275	330
Gardermoen	178	189	297	486
Sola	175	185	273	481

Table 3. Covariance (M) with respect to Liverpool, in dekameters^2

Station	1000 mb	700 mb	500 mb	300 mb
Liverpool	96	139	259	465
Camborne	80	113	209	384
Larkhill	83	122	230	415
Hemsby/Downham M.	88	127	235	420
Valentia	84	112	194	358
Aldergrove	98	138	248	438
Leuchars	105	143	259	458
Stornoway	108	136	234	421
Lerwick	100	132	228	388
Weather-ship J	57	68	86	129
Weather-ship M	77	79	131	238
Jan Mayen	18	20	41	95
Skattöra	43	36	55	114
Gardermoen	78	102	160	279
Sola	102	126	200	364

Table 4. Autocorrelation function (R) with respect to Liverpool, in per cent.

Station	1000 mb	700 mb	500 mb	300 mb
Liverpool	100	100	100	100
Camborne	90	87	89	89
Larkhill	96	94	92	92
Hemsby/Downham M.	95	94	89	91
Valentia	79	84	80	83
Aldergrove	95	95	95	94
Leuchars	91	94	94	94
Stornoway	88	84	82	81
Lerwick	78	81	78	78
Weather-ship J	46	42	32	29
Weather-ship M	59	50	50	50
Jan Mayen	15	14	18	24
Skattöra	35	23	20	29
Gardermoen	59	63	58	57
Sola	80	79	75	74

3. OBSERVATION ERRORS

The observed values of heights contain observation errors $\delta(x,y)$. Denoting the observed values by $z'(x,y)$, one has

$$z' = z + \delta .\tag{5}$$

Assuming the errors to be random and have zero mean, one finds

$$\bar{z}' = \bar{z} .\tag{6}$$

The deviations ζ' of z' from its mean value are therefore

$$\zeta' = z' - \bar{z}' = \zeta + \delta .\tag{7}$$

Under the assumption that the observation errors at the different stations are mutually independent and uncorrelated with the values of ζ, one finds that the observation errors will cause an increase of the computed variances, whereas the covariances will be unaffected by such errors.

Thus

$$N'(x_o,y_o) = \overline{\zeta'(x_o,y_o)^2} = \overline{\zeta(x_o,y_o)^2} + \overline{\delta(x_o,y_o)^2} \qquad (8)$$

$$= N(x_o,y_o) + \overline{\delta(x_o,y_o)^2}$$

whereas

$$M'(x,y,x_o,y_o) = \overline{\zeta'(x,y)\zeta'(x_o,y_o)} = \overline{\zeta(x,y)\zeta(x_o,y_o)} \qquad (9)$$

$$= M(x,y,x_o,y_o)$$

provided x,y and x_o,y_o represent different stations.

This suggests a method for determination of the mean square error. According to (9), M may be determined for several stations x,y in the vicinity of the central station x_o,y_o. From these values, $M(x_o,y_o,x_o,y_o) = N(x_o,y_o)$ may be determined by interpolation; and since $N'(x_o,y_o)$ is calculated, the mean square error $\overline{\delta(x_o,y_o)^2}$ can be obtained from (8). Alternatively, the method may be applied to the autocorrelation function instead of the covariance. In this latter case, one finds that the observation errors will cause a depression of the autocorrelations at the surrounding stations, so that the interpolated value R' for the central station is somewhat smaller than unity.

The method was applied to the autocorrelations computed for the seven stations Camborne, Larkhill, Hemsby, Valentia, Aldergrove, Leuchars and Stornoway, with respect to the central station Liverpool (compare Table 4). The interpolation was performed by fitting a second degree polynomial to the computed values by least squares. Table 5 shows the interpolated values R' of the autocorrelation at Liverpool at the four levels, and the corresponding values of $\overline{\delta^2}$.

Table 5.

Level	R'	$\overline{\delta^2}$ (meters^2)	$\sqrt{\overline{\delta^2}}$ (meters)
1000 mb	0.969	299	17.3
700 mb	0.963	516	22.7
500 mb	0.950	1295	36.0
300 mb	0.955	2092	45.7

The figures indicate a considerable increase of the errors with
height, as one would expect. It should be noted, however, that the
values obtained by this method are too large, because they are not
due to observation errors alone, but also to fluctuations in the
pressure field of scale smaller than the distance between the stations.

4. APPLICATION TO NUMERICAL ANALYSIS

Suppose that we want to determine the most probable value ζ_0'' of ζ
in a point 0 from the observed values at n near-by stations. Let
the observed values, which are subject to observation errors, be
$\zeta_1', \zeta_2', \ldots \zeta_n'$, where

$$\zeta_k' = \zeta_k + \delta_k, \qquad k = 1, 2, \ldots, n, \tag{10}$$

and let ζ_0 denote the true value in the point 0. We may then ask
for the linear formula

$$\zeta_0'' = \sum_1^n a_k \zeta_k', \qquad a_k \text{ constants}, \tag{11}$$

which makes

$$U = \overline{(\zeta_0'' - \zeta_0)^2} = \text{a minimum.} \tag{12}$$

Inserting (11) into (12) and differentiating with respect to a_j, one
obtains

$$\overline{\zeta_j' (\sum_1^n a_k \zeta_k' - \zeta_0)} = 0, \qquad j = 1, 2, \ldots, n,$$

or

$$\sum_1^n \overline{\zeta_j' \zeta_k'} \, a_k = \overline{\zeta_j' \zeta_0}, \qquad j = 1, 2, \ldots, n, \tag{13}$$

which is a linear system of equations for determination of the constants a_k. The coefficients of these equations are seen to be covariants of the type discussed above. With the notation $\overline{\zeta_j \zeta_k} = M_{jk}$, $\overline{\zeta_j' \zeta_k'} = M_{jk}'$, $\overline{\zeta_j'^2} = M_{jj}' = N_j'$, $\overline{\zeta_j^2} = M_{jj} - N_j$, and $\overline{\zeta_j' \zeta_0} = M_j$, one has

$$M_{jk}' = M_{jk} \quad (j \neq k), \qquad M_{jj}' = M_{jj} + \overline{\delta_j^2}. \tag{14}$$

The system (13) may therefore be written

$$\sum_{k=1}^{n} M'_{jk} a_k = M_{jo} \tag{15}$$

or

$$\sum_{k=1}^{n} M_{jk} a_k + \overline{\delta_j^2} \, a_j = M_{jo} \tag{16}$$

It is easy to see that the determinant of this system does not vanish. For any set of numbers $u_1, u_2, \ldots u_n$, which are not all zero,

$$\overline{(\sum_{k=1}^{n} \zeta'_k u_k)^2} = \sum_{j=1}^{n} \sum_{k=1}^{n} \overline{\zeta'_j \zeta'_k} \, u_j u_k = \sum_{j=1}^{n} \sum_{k=1}^{n} M'_{jk} u_j u_k > 0.$$

This quadratic form in the u's is thus positive definite, and consequently

$$M'_{11} > 0, \quad \begin{vmatrix} M'_{11} & M'_{12} \\ M'_{21} & M'_{22} \end{vmatrix} > 0, \quad \ldots \quad \begin{vmatrix} M'_{11} & M'_{12} & \cdots & M'_{1n} \\ M'_{21} & M'_{22} & \cdots & M'_{2n} \\ \vdots & & & \\ M'_{n1} & M'_{n2} & & M'_{nn} \end{vmatrix} > 0 \tag{17}$$

The last of these equations shows that the determinant of the system (15) is positive.

The n equations (17) are conditions which the covariances must fulfill. The writer is indebted to Mr. J. Nordö for the remark that these conditions are equivalent with the conditions that the partial correlation coefficients of all orders must be numerically less than one.

The minimum value of U (eq. (12)) is found to be

$$U_{min} = M_{oo} - \sum_{k=1}^{n} a_k M_{ko} \tag{18}$$

which is a measure of the quality of the "analysis" formula (11).

When this method is applied to obtain the values of ζ'' in a sufficient number of points between the observing stations, one obtains a numerical analysis of the field. It is easy to see that if $\overline{\delta^2} = 0$, the resulting analysis will represent a pure interpolation between the observed values; but if $\overline{\delta^2} \neq 0$, one obtains a combined interpolation and smoothing, and the degree of smoothing will depend upon the ratio of the values of $\overline{\delta^2}$ to the values of the variances N.

5. ANALYSIS OF THE SPECTRUM OF PRESSURE SYSTEMS OF DIFFERENT SCALE

Consider a rectangular region with side lengths A and B in the xy-plane. A two-dimensional Fourier expansion of $\zeta(x,y)$ over this region may be written

$$\zeta(x,y) = \sum_{m,n=-\infty}^{+\infty} c_{m,n}\, e^{2\pi i (m\frac{x}{A} + n\frac{y}{B})} \tag{19}$$

where $c_{m,n}$ are complex constants. In order that ζ be real, one must have that

$$c_{-m,-n} = c_{m,n}^{*} \tag{20}$$

where the star denotes conjugate complex value.

Let brackets denote space average over the rectangular region. A spatial covariance may be defined as

$$Q(\xi,\eta) = [\zeta(x,y)\, \zeta(x+\xi, y+\eta)] \tag{21}$$

By Parseval's theorem, one has

$$Q(\xi,\eta) = \sum_{m,n=-\infty}^{+\infty} c_{m,n}\, c_{-m,-n}\, e^{2\pi i (m\frac{\xi}{A} + n\frac{\eta}{B})}$$

$$= 2 \sum_{m,n=0}^{+\infty} \{(c_{m,n}\, c_{-m,-n} + c_{m,-n}\, c_{-m,n})\cos(2\pi m \tfrac{\xi}{A})\cos(2\pi n \tfrac{\eta}{B}) \tag{22}$$

$$- (c_{m,n}\, c_{-m,-n} - c_{m,-n}\, c_{-m,n})\sin(2\pi m \tfrac{\xi}{A})\sin(2\pi n \tfrac{\eta}{B})\}$$

Hence it follows, when $\xi = \eta = 0$,

$$Q(0,0) = [\zeta^2] = 2 \sum_{m,n=0}^{\infty} (c_{m,n} c_{-m,-n} + c_{m,-n} c_{-m,n}) \quad (23)$$

showing that the quantity

$$e_{m,n} = c_{m,n} c_{-m,-n} + c_{m,-n} c_{-m,n} \quad (24)$$

represents the "energy" $[\tfrac{1}{2}\zeta^2]$ of the Fourier component with wave numbers m,n.

From (24) and (22) one finds

$$e_{m,n} = \frac{2}{AB} \int_0^A \int_0^B Q(\xi,\eta) \cos(2\pi m \tfrac{\xi}{A}) \cos(2\pi n \tfrac{\eta}{B}) \, d\xi \, d\eta, \quad (25)$$

which shows the well-known connection between the energy spectrum and covariance.

It is of course not quite correct to apply this analysis to horizontal fields on a spherical earth. The proper representation would be spherical harmonics, rather than Fourier analysis, and the area of integration would have to be extended to the entire globe, so that the result would represent the mean energy distribution over the earth. However, considering that the statistical properties of the pressure field are widely different in different parts of the world, it would seem desirable to define the spectrum of energy more locally for limited regions, at least for wave lengths considerably shorter than the circumference of the earth. It is reasonable to assume that equation (25) could be used for this purpose, provided the sides A and B of the basic rectangle are chosen considerably smaller than the earth's radius.

It will be noted that the subdivision of the mean "energy" $[\tfrac{1}{2}\zeta^2]$ into a discrete spectrum according to (23) is somewhat arbitrary, because the discrete set of wave lengths corresponding to integral numbers m,n depends upon the arbitrary assignment of A and B. Thus, for instance, the energy of pressure systems of dimensions larger than A and B would erroneously be attributed to waves with wave lengths equal to or smaller than A and B.

Instead of asking for the energy of these fictitious spectral lines, it is perhaps more reasonable to find the amount of energy contained within a given range of wave lengths. Introducing

$$k_x = \frac{2\pi m}{A} \quad , \quad k_y = \frac{2\pi n}{B} \quad , \tag{26}$$

and noting that $e_{m,n}$ represents the energy in a wave number interval

$$\Delta k_x = \frac{2\pi}{A} \quad , \quad \Delta k_y = \frac{2\pi}{B} \quad , \tag{27}$$

one may define the quantity

$$E(k_x, k_y) = \frac{e_{m,n}}{\Delta k_x \Delta k_y} = \frac{1}{2\pi^2} \int_0^A \int_0^B Q(\xi,\eta) \cos k_x \xi \cos k_y \eta \, d\xi \, d\eta \tag{28}$$

which represents the spectral energy distribution without reference to the fictitious spectral lines. Since this equation is derived from (25), it applies only to wave lengths in the x and y-directions which do not exceed A and B respectively.

It will be seen from the form of the right-hand side of (28), since Q is a damped function, that E will be almost independent of A and B for sufficiently short wave lengths; in this case, (28) may be replaced by

$$E(k_x, k_y) = \frac{1}{2\pi^2} \int_0^\infty \int_0^\infty Q(\xi,\eta) \cos k_x \xi \cos k_y \eta \, d\xi \, d\eta \quad . \tag{29}$$

Thus, for sufficiently short waves, the function E defined by (28) would have a definite meaning. For longer waves (but shorter than A and B), E defined by (28) would depend upon A and B and would therefore be more or less arbitrary. In particular, E would contain energy which should properly have been connected with systems of wave lengths larger than A and B. This difficulty is avoided in the theory of homogeneous turbulence, where A and B may be chosen sufficiently large so that (29) holds for the entire spectrum considered. This cannot be done in the present case; owing to the statistical non-homogeneity of the pressure field, the rectangle AB must be chosen so small that the statistical properties may be considered as approximately homogeneous within this region. Therefore, (28) can be used only for the short-wave end of the spectrum, where the convergence of the integral is fast enough so that the result is independent of A and B.

Time averaging of (28) gives

$$\overline{E(k_x, k_y)} = \frac{1}{2\pi^2} \int_0^A \int_0^B \overline{Q(\xi,\eta)} \cos k_x \xi \cos k_y \eta \, d\xi \, d\eta \quad . \tag{30}$$

On the other hand one has on account of (21) and (3)

$$\overline{Q(\xi,\eta)} = \overline{[\zeta(x,y) \, \zeta(x + \xi, y + \eta)]} = M(x, y, x + \xi, y + \eta) \quad . \tag{31}$$

These equations indicate that it may be possible to calculate the mean energy distribution function on the basis of calculations of the spatial covariance M.

If one is justified in assuming that, for a limited region, $M(x,y, x + \xi, y + \eta)$ is primarily a function of ξ and η, then $\overline{Q}$ in equation (30) could simply be replaced by $M(x_0, y_0, x_0 + \xi, y_0 + \eta)$. This may not be a good approach, however, because the latter function has its maximum at a considerable distance from the point $\xi = 0$, $\eta = 0$, (compare Table 3), whereas $Q(\xi,\eta)$ is strictly symmetric with respect to this point. It would perhaps be better to replace $\overline{Q(\xi,\eta)}$ by

$$[N] \cdot R(x_0, y_0, x_0 + \xi, y_0 + \eta) \quad , \tag{32}$$

since this function is fairly symmetric with respect to the central point. It is the author's intention to make an attempt to compute the mean energy distribution function in this manner.

Acknowledgement. The author wishes to thank Dr. E. Höiland and Mr. J. Nordö for valuable discussions.